KV-372-986

MONOGRAPHS OF THE PHYSIOLOGICAL SOCIETY

Editors: H. Davson, A. D. M. Greenfield,
R. Whittam, G. S. Brindley

Number 21 BIOGENESIS AND PHYSIOLOGY OF
HISTAMINE

Volumes marked * are now out of print.

BIOGENESIS AND PHYSIOLOGY
OF HISTAMINE

GEORG KAHLSON   and   ELSA ROSENGREN

M.D., LL.D.ST.AND.   M.D.

Professor of Physiology   *Lecturer in Physiology*

Institute of Physiology,
University of Lund, Sweden

MAGDALEN LIBRARY COLLEGE

LONDON
EDWARD ARNOLD (PUBLISHERS) LTD

© GEORG KAHLSON and ELSA ROSENGREN, 1971

First published 1971
by Edward Arnold (Publishers) Ltd.,
41 Maddox Street,
London, W1R 0AN

ISBN 0 7131 4184 0

All Rights Reserved. No part of this publication may
be reproduced, stored in a retrieval system, or trans-
mitted in any form or by any means, electronic mech-
anical, photocopying, recording or otherwise, without
the prior permission of Edward Arnold (Publishers)
Ltd.

Printed in Great Britain by
The Camelot Press Ltd, London and Southampton

PREFACE

THE invitation by the Monograph Committee of the British Physiological Society to contribute to their series of special publications I accepted with enthusiasm and a sense of encouragement. The assignment by the Committee was for a Monograph on the work on histamine done in my laboratory. Having worked in the years 1937–1938 under Jack Gaddum at University College, London, and under Henry Dale at the National Institute for Medical Research, I embarked on this task with devotion and humility towards these teachers and leaders in histamine research.

Although the assignment by the Monograph Committee was a limited one, confined to work in this field done in my laboratory, I soon became aware that writing even this little book alone was a task beyond my ability. And so I asked my co-worker for many years, Dr. Elsa Rosengren, to join the venture. This association is appropriate since what follows is an account of work on histamine carried out by many members of the Family of Physiologists of Lund over a period of about thirty years.

In order to be useful to readers, our account has been placed in proper context with observations by other workers in the field. This we have tried to do. Yet, aiming at a brief and straightforward presentation of our work and views, such references of necessity are few. Readers seeking information on the wide realm of histamine physiology will find our account narrow and unsatisfactory. Histamine is covered comprehensively in the *Handbook on Histamine*, Volume I, edited by Rocha e Silva, with 991 pages by thirty-seven authors published in 1966. Volume II is to follow.

This Monograph has been written in the kind of English at our command, and any blame for dullness in style is not to be laid on a translator.

GEORG KAHLSON

ACKNOWLEDGEMENTS

OUR gratitude goes to the Editors, Dr. Hugh Davson who read the first eight chapters, and to Professor A. D. M. Greenfield who read the whole manuscript, making suggestions for improvements towards greater clarity and smoother style.

We are grateful to Miss Barbara Koster of Arnolds for her untiring help and collaboration in the preparation of the text, references and figures. Our laboratory assistant Miss Lena Olsson typed, retyped, kept order and had to live with preparations for the book for two years.

Acknowledgement for permission to reproduce published figures is due to the editors and publishers of the following journals:

Acta chirurgica scandinavica; *Acta medica scandinavica*; *Acta physiologica scandinavica*; *Biochemical Journal*; *Biochemical Pharmacology*; *British Journal of Pharmacology*; *Experientia*; *Experimental Cell Research*; Academic Press, New York, *Federation Proceedings*; *Gastroenterology* (Williams & Williams, Baltimore, 1967); *Immunology*; *Journal of Histochemistry and Cytochemistry*; *Journal of Physiology*; *Lancet*; *Life Sciences*; *Medicina experimentalis*; *Nature*; *Perspectives in Biology and Medicine*; *Physiological Reviews*; *Proceedings of the National Academy of Sciences, Washington*; *Proceedings of the Society of Experimental Biology and Medicine*; *Scandinavian Journal of Clinical and Laboratory Investigation*.

CONTENTS

1

INTRODUCTION

HISTAMINE had a flying start in the hands of pioneers, Henry Dale, Thomas Lewis, Eugen Werle, Peter Holtz, Carl Dragstedt, Wilhelm Feldberg, Jack Gaddum, and others. Here was a body constituent seemingly designed to play an essential part in the machinery of the body. Yet, in spite of persistent search for decades, by workers all over the world, no place could be found for histamine in normal physiology. Indeed, in 1950, at a Symposium at the Royal Society of London, histamine was relegated to the sphere of pathology, designated as 'not strictly physiological, and liberated when cells are injured' (Burn, 1950). Even at the sumptuous *Symposium on Histamine* in honour of Sir Henry Dale, held in London in 1955, there was little material to portend the imminent watershed in histamine research. On that occasion the biogenesis of histamine and its release, and the association of the amine with mast cells were the dominant topics.

The mast cell came to the fore by Riley's discoveries, beginning in 1952. The particular bearing of this work has been in the words of Riley and West (1966): 'Through the discovery of the high histamine content of the mast cells of Ehrlich, the decade 1950–1960 has seen a fresh orientation of knowledge concerning the location of histamine in blood and tissues.' The discovery by MacIntosh and Paton, in 1949, of what came to be known as histamine liberators, a wide range of organic bases, intensely explored after the finding that these compounds release histamine from mast cells, provided the background for new kinds of experiments on histamine. Here was a handy model, it was then believed, for studying, at the cellular level, the mechanism of histamine release in anaphylaxis. Part of the immense amount of work done on histamine release has been reviewed in the *Handbook on Histamine* edited by Rocha e Silva, published in 1966.

High hopes were at first entertained that the antihistamines, discovered in France about thirty years ago, would provide a master-key, as it were, for detecting the involvement of histamine in physiological phenomena. The antihistamines have, however, proved to be of little or no avail in this search, because these compounds, as it will be explained, do not antagonize the dynamic process of intracellular histamine formation, which is the means by which histamine becomes part of physiological events.

Henry Dale, in 1962, looking back on events, emphasized the 'anomalies and difficulties, the paradoxes and conundrums, with which the investigator has found himself called upon to deal, in studies of the presumptive participation of histamine in normal and pathological reactions'. Really, explorers of the physiology of histamine have had to tread a devious path. Until about 1960 it was believed that alterations in the histamine content of tissues, the release therefrom, and its excretion, would reflect the participation of histamine in normal reactions, and, in fact, other approaches were not practicable owing to limitations in the methods then available. In the late fifties, Schayer in the U.S.A., and physiologists at the University of Lund, independently showed that in normal and pathological reactions great changes in the rate of endogenous histamine formation did occur without any corresponding changes in tissue histamine content. Furthermore, means were contrived to inhibit or enhance the rate of histamine formation *in vivo*. The concepts of what we refer to as 'non-mast-cell histamine' and 'histamine forming capacity' have readily become accepted. Histamine research took a determined course, guided by new views and adequate techniques, among which Schayer's isotopic methods have been dominant. How, and why, events came to change the course, and what new lands have come in sight, is the theme of this book, which perhaps should be read straight through like the tale of a journey.

2

TERMINOLOGY

A NUMBER of terms and definitions used in reports from our laboratory are listed below and their context explained. Three among these are novel: histamine forming capacity (HFC), non-mast-cell histamine, and nascent histamine.

Endogenous histamine: Histamine (Hi) formed within tissue cells by decarboxylation of histidine, predominantly by the agency of histidine decarboxylase.

Exogenous histamine: Hi from sources other than tissue cells, e.g. intestinal contents, injected Hi.

Histidine decarboxylase: intracellular enzyme, with high affinity for histidine, instrumental in the formation of endogenous Hi. The prefix 'specific' appears superfluous. Decarboxylases with low affinity for histidine, currently given the prefix 'non-specific', should preferably not be designated as histidine decarboxylase. Characteristics of the enzyme are described in Chapter 6.

Histamine forming capacity (HFC): rate at which Hi is formed in minced tissues, cell suspensions, or the whole body; it denotes histidine decarboxylase activity in numerical terms (Kahlson, Rosengren & White, 1960).

Non-mast-cell histamine: Hi formed in cells other than mast cells; in non-mast cells the HFC can be strikingly high and the resulting Hi is often 'nascent' in nature. The term was introduced in 1960 (Kahlson, Rosengren & White, 1960) and the distinction between mast cell and non-mast-cell histamine was explored in 1963 (Kahlson, Rosengren & Thunberg, 1963).

Nascent histamine: a term introduced 1962 (Kahlson). Intracellular Hi, formed at very high rates in some tissues and circumstances, presumably non-mast-cell Hi, seemingly involved in certain kinds of rapid tissue growth. Its action appears to be linked to the very process of its formation; not to be confused with the liberation of preformed Hi, notably not with the Hi liberated and acting 'intrinsically', as envisaged by Dale (1948). The action of nascent Hi is presumably not achieved by injected Hi, and its action is not prevented by antihistaminics.

Liberation, release, mobilization of Hi: implies transfer of preformed Hi from intracellular to extracellular state; this has in normal physiology so far been shown to occur in one instance only, in the gastric mucosa on feeding.

3

ORIGIN OF THE HISTAMINE CONTAINED IN TISSUES

'THE origin of histamine in the body' was one of the principal themes for discussion at the symposium on histamine, held in London in 1955, in honour of Henry Dale, published as *Ciba Symposium on Histamine* (1956). At that time results had been obtained which seemingly showed that histamine forming enzymes were not present in any of the tissues of cats and dogs. In these animals the tissue histamine was believed to be absorbed from the gut, thus acquiring, as it were, a vitamin-like status. It had further been reported that treatment of rats with compounds that inhibit the growth of intestinal bacteria was accompanied by a substantial decrease in urinary histamine excretion and tissue content of the amine. That is to say, in some species tissue histamine were wholly of exogenous origin, whereas in other species, such as the rat, the tissue supply of histamine was dual in origin, at least in part endogenous. The methods and experiments on which this view was based will be discussed later. If this view were correct, histamine of endogenous origin would be swamped by exogenous histamine and determination of histamine content in tissues and urine would be of little physiological significance.

In the course of work with quite different objects, we incidentally made observations which suggested that histamine contained in tissues is also formed therein.

Observations on starved cats. Complete withdrawal of food, except for water, for as long as seven days did not significantly alter the content of histamine in the mucosa of the stomach, the small intestine, the colon or the rectum in adult cats. In this experiment the intestinal content of histidine available for bacterial decarboxylation must have been very low and, equally low, the amount of histamine to be absorbed. Nonetheless, the mucosal histamine content did not fall, which suggested that the level of histamine in the gastro-intestinal mucosa is fundamentally controlled by factors other than those prevailing in the intestine (Haeger & Kahlson, 1952*a*). This modest beginning led to further experiments from which it would appear, perhaps, that these levels of intestinal histamine may be genetically determined. The observed constancy in histamine content is the more remarkable as in the gastro-intestinal mucosa its histamine is not, or is only negligibly, contained in mast cells.

'Regulatory mechanism' controlling the levels of histamine. Having seen that the levels of histamine in the mucosa of the gastro-intestinal tract remain the same independent of any amount of the amine presumed to be absorbed from the gut, in order to clarify the situation further, the following experiments were designed (Haeger, Kahlson & Westling, 1953). In cats at different stages of gestation, and in each individual foetus, the stomach and small intestine were removed and their histamine concentrations determined. Since it was difficult in the foetus to separate the mucosa from the other layers the whole wall of the stomach and small intestine were investigated in the foetus and also in the mother in which ten segments 0·5–1 cm in length were cut out from the small intestine and pooled, whereas in the foetus the entire small intestine was extracted.

The values obtained for the stomach in eight cats and the corresponding foetuses are given in Fig. 1. In the gastric wall of the mothers the histamine concentration differs greatly within the group, being spread at random between the extremes 4 and 43 μg/g. Now consider the individual mother and the corresponding foetuses: in the mother and her foetuses the histamine concentration of the gastric wall is approximately the same. There is one exception (Nr. 1) in which the concentration in the mother is much higher than in the foetuses, among which, again, the level is the same.

In the intestinal wall of these mothers the histamine concentration falls within approximately the same range as in their stomach wall, 11–46 μg/g in the intestine, as seen in Fig. 2. In contrast to the levels prevailing in the stomach, the histamine concentration of the foetal small intestine is throughout lower than in the corres-

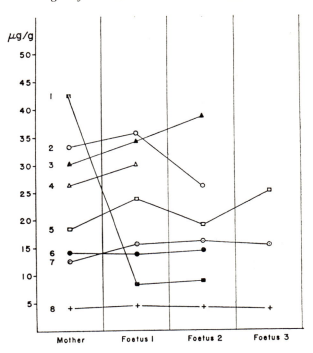

FIG. 1. Histamine concentration (μg histamine base per gram) of the whole stomach wall in eight mothers and corresponding foetuses.

Weight of foetuses in grams:

Cat no: 1: 100 ± 5 2: 32 ± 2 3: 100 ± 5
       4: 40 ± 2 5: 85 ± 4 6: 150 ± 5
       7: 135 ± 5 8: 110 ± 5.

(Haeger *et al.*, 1953, *Acta physiol. scand.* **30**, Suppl. 111, 177).

ponding mother. Nevertheless, in the intestinal wall, the histamine concentration is approximately the same in foetuses of the same mother. The higher values found in the mother are not likely to be due to uptake of histamine from the intestinal tract, as has been mentioned in the preceding paragraph. The uniformity of the intestinal histamine level in foetuses of the same mother is likely to be governed by the same circumstances which attain uniformity in the foetal stomachs of the same mother.

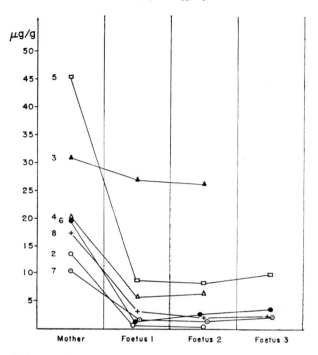

FIG. 2. Histamine concentration of the whole wall of the small intestine in the same animals as in Fig. 1 (Haeger *et al.*, 1953, *Acta physiol. scand.* **30**, Suppl. 111, 177).

The sole element common to the foetuses and the mother is the blood flowing through the umbilical vessels. After birth this uniting factor is lost and the kittens are free to establish their individual histamine levels. This, indeed, they do, and attain individual histamine levels which are different not merely from the mother, but also different from their litter mates. The pertinent results regarding the histamine content of the gastric mucosa are shown in Figs. 1 and 3 which comprise mothers, foetuses and young. The uniformity in levels which exists before birth is now lost, the values being scattered at random within the wide range of the species. The individual ways taken by the kittens is depicted in Fig. 3 which contrasts with the uniformity seen in Fig. 1.

It is noteworthy that in the cat the histamine concentration of the gastric mucosa, even during the foetal stage of development,

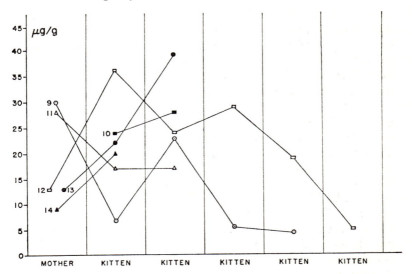

FIG. 3. Histamine concentration of the gastric mucosa of five mothers and their litters. The sample from one mother was lost (No. 10) (Haeger *et al.*, 1953, *Acta physiol. scand.* **30**, Suppl. 111, 177).

reaches the level at which it remains in the mother. The situation in the small intestine is different, a contrast which may be taken to indicate a difference in the function of gastric and intestinal histamine.

With regard to the 'regulatory mechanism', this would appear to be either blood-borne or the expression of genetic determination. Whatever alternative may be correct, these experiments appear to preclude exogenous histamine as a factor of any consequence in establishing the histamine levels of the gastro-intestinal tract.

Canadian investigators determined the histamine content of various tissues in fed and fasting (48 hr) mice, rats, hamsters and guinea-pigs. They found no evidence that feeding, after a prolonged fast, affected the histamine content of blood, lung, liver, stomach, small intestine and large intestine (Kowalewski, Russell & Koheil, 1969). That is to say, exogenous histamine did not add to the tissue histamine levels of these rodents.

Observations in germ-free reared rats. This investigation was undertaken with the object of establishing (1) whether germ-free rats differ from ordinary rats in the histamine content of their various tissues, (2) whether the urinary excretion in ordinary rats, as in the germ-free ones, reliably indicates the rate of endogenous whole-body histamine formation, and, by implication, physiologically occurring alterations in the rate of histamine formation.

At the time of this investigation (Gustafsson, Kahlson & Rosengren, 1957) germ-free rats were reared in small number by Dr. Gustafsson at the Department of Histology, University of Lund. The germ-free as well as the ordinary rats investigated were offspring of a strain of rats reared for many years at the Histology Department. After weaning, the rats were fed a special diet and water *ad lib.* The composition of the diet, which was rich in casein, is seen in Table 1. The diet, mixed with water, was autoclaved at 120°C. Of this diet, which was histamine free (<0.0024 µg/g), the rats ate about 15 g daily. Each rat was kept in a metabolism cage, and the urine and faeces were collected separately in 24-hr samples. Sterility tests were performed twice a week on faeces and waste from the germ-free apparatus.

Table 1 : Composition of the diet

Main constituents (g/kg diet)		Vitamins (mg/kg diet)		Vitamins (mg/kg diet)	
Casein	220	Vitamin A	21,000 i.u.	Nicotinamide	200
Wheat starch	630	Vitamin D	4500 i.u.	Choline	2000
Arachis oil	100	Vitamin E	500	Inositol	1000
Salt mixture*	40	Vitamin K_1	10	*p*-Aminobenzoic	
Vitamin		Thiamine	50	acid	300
mixture	10	Riboflavin	20	Biotin	1
		Pyridoxin	20	Folic acid	20
		Calcium		Vitamin B_{12}	0·02
		panto-		Ascorbic acid	1000
		thenate	100		

* The salt mixture was the same as described by Hubbell, Mendel & Wakeman, 1937, *J. Nutr.* **14**, 273.

(Kahlson, Rosengren & Westling, 1958, *J. Physiol.* **143**, 91.)

Beside various individual tissues, whole rats were extracted for histamine with the exclusion of the bony parts.

Mast cells were counted in the mesentery, and no difference in their number was found between germ-free and ordinary rats.

Table 2. Distribution of histamine in three groups of rats. Group 1: germ-free, group 2: not-germ-free on sterilized food, group 3: not-germ-free on not sterilized food. The figures represent μg histamine base per gram tissue

No. age and sex	Ear mean of two	Skin of the back	Lung	Small intest.	Gastric mucosa	Gastrocnem. muscle	Tongue
Group 1							
ca. 120 days old							
1 ♂	23·7	10·6	8·7	*—	—	1·7	45·8
2 ♂	26·4	12·3	13·8	37·1	—	3·9	49·5
3 ♂	25·8	11·9⎫ †	7·8	12·3	30·8	4·9	27·8
4 ♂	22·0	13·9⎬	6·9	9·1	16·2	4·1	23·5
5 ♀	19·5	11·5⎭	5·1	10·8	56·4	4·8	30·5
ca. 300 days old							
6 ♀	16·3⎫	3·6	8·0	21·0	41·4	5·6	33·3
7 ♀	18·6⎬ ‡ 9·8		6·0	16·0	47·6	—	26·1
8 ♂	20·2⎭	6·3	6·1	—	35·7	4·4	35·5
Group 2							
ca. 120 days old							
9 ♂	26·4	10·5	2·1	24·4	40·7	3·0	28·5
10 ♀	25·0	28·5	3·3	23·8	63·5	3·4	43·6
11 ♀	19·9	22·0	8·2	21·7	58·9	3·1	36·3
12 ♂	24·0	10·9	5·8	40·8	43·3	2·7	—
13 ♂	26·1	21·1	2·1	10·6	68·7	3·4	—
14 ♀	23·4	20·2	5·6	33·2	61·5	4·3	—
Group 3							
ca. 120 days old							
15 ♂	29·5	23·1	7·3	20·3	33·0	2·3	—
16 ♂	25·5	16·6	8·0	22·1	51·5	3·2	—
17 ♂	28·9	25·2	4·3	9·4	28·9	1·9	—
18 ♂	30·0	24·0	6·0	28·5	53·9	2·8	—
19 ♂	33·4	19·6	3·5	15·6	28·5	2·0	—
20 ♀	31·3	25·9	4·0	16·8	—	3·1	—
ca. 300 days old							
21 ♂	9·7	6·4	—	20·0	—	—	30·8
22 ♂	9·5	6·4	—	22·9	—	—	32·5
23 ♂	11·7	10·9	—	48·2	—	—	45·4

*— stands for sample not investigated or lost.

† Mean of two specimens of skin.

‡ One ear only.

(Gustafsson *et al.*, 1957, *Acta physiol. scand.* **41**, 217.)

The histamine content of various organs in three groups of rats was investigated (Table 2). Group 1 comprises eight germ-free rats of both sexes, five about 120 days old and three about 300 days. Group 2 consists of six non-germ-free rats of both sexes, about 120 days old and fed the sterilized synthetic diet. Group 3 represents nine non-germ-free rats of both sexes, six about 120 days old and three about 300 days old, fed the non-sterilized synthetic diet.

Because of the very great variations among individuals in the histamine levels in the skin seen in Table 2, this tissue was reinvestigated in two groups, germ-free and non-germ-free, where in addition the histamine content of the whole rats was determined. The results are summarized in Table 3.

Table 3. Histamine content in μg base per gram tissue and total histamine in the whole animal in germ-free and not-germ-free rats

No. sex and age in days	Weight (g)	Skin of the back	Abdominal skin	Total skin	Tongue	Soft tissue	Total histamine whole animal	Histamine per gram whole animal
Germ-free								
1 ♀ 201	200	—	—	11·8	—	7·7	1378	6·9
2 ♂ 95	270	18·8	26·4	22·9	—	5·7	1770	6·6
3 ♀ 166	210	10·2	17·5	13·6	45·8	—	—	—
4 ♂ 104	230	15·8	22·7	19·9	49·5	8·3	1795	7·8
Not germ-free								
5 ♂ 106	235	14·3	22·2	18·2	28·5	3·4	1163	5·0
6 ♂ 168	370	10·0	—	11·2	43·6	—	—	—
7 ♀ 125	220	—	—	12·8	36·3	—	—	—
8 ♀ 117	175	—	—	12·9	—	4·8	886	5·1
9 ♂ 117	255	—	—	15·1	—	3·9	1219	4·8
10 ♂ 165	360	—	19·0	17·1	—	7·9	2635	7·3
11 ♂ 149	350	12·7	16·4	13·6	—	5·0	1858	5·3

(Gustafsson *et al.*, 1957, *Acta physiol scand.* **41**, 217.)

The values summarized in Tables 2 and 3 do not easily lend themselves to statistical analysis. However, it appears certain that the germ-free rats, neither in any particular tissue examined, nor in the whole animal, contain less histamine than the non-germ-free ones. Indeed, the uniformity in tissue histamine content between the groups should be taken as conclusive evidence that even in non-germ-free rats fed on a histamine free diet, the histamine contained in the various tissues is endogenous in origin.

In the course of this work startling observations were made regarding the urinary excretion of histamine. In these experiments urinary free histamine was assayed directly on a segment of guinea-pig ileum, as previously done by Wilson (1954). The hydrolysis of conjugated urinary histamine was carried out by boiling with concentrated HCl. A sex difference in histamine excretion was noted which was unexpected at the time of these experiments. The findings in males will first be described. Fig. 4 represents observations on three germ-free and three non-germ-free rats. In the germ-free group the lowest and highest amounts of total histamine excreted in 24 hr are 9 and 23 μg, respectively. In the non-germ-free rats the corresponding figures are 6 and 17 μg. These values were independent of the urine volumes. In the non-germ-free group nearly all the histamine is excreted in the conjugated form. The germ-free males excrete free histamine in considerable amounts, its proportion of the total excretion appears subjected to considerable variations between individual rats.

Only one germ-free female could be investigated closely because of the small amount of material available. Numerous non-germ-free females were investigated of which two typical ones are included in Fig. 5. In contrast to the male, nearly all the histamine excreted is in the free form. The germ-free female excreted large amounts of both free and conjugated histamine. At that time it had been shown that ordinary rats can acetylate histamine and excrete acetylhistamine [4-(β-acetylaminoethyl) imidazole]; this work has been reviewed by Tabor (1954, 1956). In order to see whether the conjugate was in fact acetylhistamine, a total of 65 ml urine was collected from a germ-free female rat and subjected to a series of procedures, as described in our original paper. Finally, on running over No. 1 Whatman paper, authentic acetylhistamine and the purified urinary conjugate advanced at the same rate. To our knowledge, no further attempts have been made to identify the chemical nature of the conjugated histamine excreted in germ-free female rats. On the evidence of observations in man, Sjaastad (1967) called for a reappraisal of the hitherto supposed unimportant role allotted to acetylation in the catabolism of histamine.

It should be added that the faeces of the various groups of rats examined in this study did not contain free or conjugated histamine in measurable quantities.

The principal outcome of this work is the recognition that the

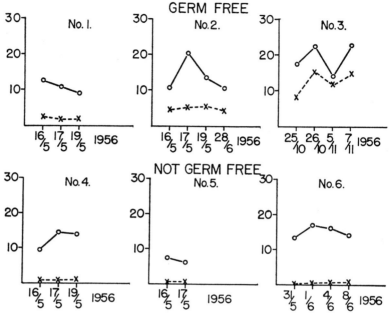

FIG. 4. Urinary excretion of histamine in μg/24 hr in three germ-free and three non-germ-free male rats, all fed a histamine-free sterile diet. x-x-x free histamine, o—o—o total histamine (Gustafsson *et al.*, 1957, *Acta physiol. scand.* **41**, 217).

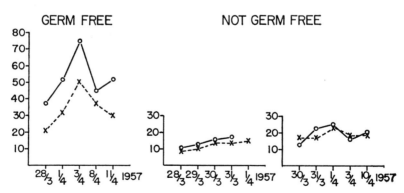

FIG. 5. Urinary excretion of histamine in μg/24 hr in one germ free and two not-germ-free female rats. x - - - x - - - x free histamine, o—o—o total histamine (Gustafsson *et al.*, 1957, *Acta physiol. scand.* **41**, 217).

whole-body endogenous histamine formation and changes therein can be followed continuously day after day in the female rat by observations of the urinary excretion of histamine. The first fruit to be gathered by this means was the discovery that in the pregnant rat histamine is formed at exceedingly high rates. The female rat and her urinary excretion of histamine is now widely used to reveal changes in endogenous histamine formation.

In retrospect, the basic principle of the above approach for measuring changes in body histamine was forestalled by Gaddum's pronouncement (1951): 'The estimation of free histamine in the urine is thus the most sensitive way we know of detecting the liberation of histamine in the body.' At that time histamine liberation, not formation, was in the fore.

Determinations of tissue histamine content in germ-free and regular rats have subsequently been carried out by Levine, Sato and Sjoerdsma (1965), who confirmed the similarity between the two groups with respect to tissue histamine content. Nevertheless, from observations with alleged *in vivo* inhibitors of histidine decarboxylase these authors stated, 'a significant proportion of the histamine remaining in the urine was probably formed by intestinal bacteria'. In the experiments of Levine *et al.*, the histamine excretion in untreated germ-free rats was 64% lower than in regular rats, whereas the histamine levels in heart, stomach and mast cells were the same in the two kinds of rat.

On mono-associating germ-free rats with a histamine-producing strain of *Clostridium perfringens*, Beaver and Wostmann (1962) found high histamine concentrations in the caecal contents, but the histamine levels in the wall of the small intestine did not increase. In untreated germ-free rats these authors found the concentration of histamine in the wall of the caecum slightly higher than in regular rats, whereas in the intestinal wall the histamine levels were equal in two investigated series, but in a third series the histamine concentration in the wall of the small intestine was slightly higher in regular than in germ-free rats.

In a study of histamine excretion as influenced by X-ray irradiation and compound 48/80, Leitch, Debley and Haley (1956) made observations which are pertinent to this topic. They submitted rats to sterilization of the gut by daily administration of 50 mg/kg of sulfasuxadine for 59 days and 160 mg/kg of terramycin for 8 days. This drastic treatment did not diminish the urinary excretion of histamine. In the course of this work Leitch *et al.* found, incidentally as we did, that the female rat excretes much more free histamine than the male.

In sum, in the rat the evidence appears overwhelming, and in the cat circumstantial, in support of the view that the histamine contained in tissues is also formed therein. Besides, to students of the physiological roles of histamine it appears by and large rather irrelevant whether in some species, in certain tissues, under certain experimental conditions, small amounts of histamine can be taken up from the outside of tissue cells.

4

UPTAKE OF HISTAMINE FROM THE OUTSIDE

In the various species so far studied, each tissue examined forms histamine by decarboxylation of histidine, although at exceedingly varying rates. There is no formal relation, however, between the rate of formation and histamine content in tissues. It will be explained later in some detail that the histamine content of a tissue to a large extent depends on the life-time and mode of binding of histamine which are characteristic of individual tissues. Owing to the limited intracellular life-time, histamine is continuously leaving the tissues into the blood stream to be excreted in the urine, either unchanged or catabolized. That is to say, a steady stream of blood-borne histamine exists, even under entirely physiological conditions.

It is known that tissue cells take up exogenous histamine under certain experimental conditions and some investigators maintain that this process of uptake is of importance in providing for adequate stores of histamine. A few investigations only will be cited in this connexion. Waton (1964) infused ^{14}C-histamine in cats and found a fraction present in all the tissues examined 90 hours after the infusion. He believes that in the cat histamine 'first of all might be formed in the tissue where it is found or it may be formed in another tissue and transported by the blood or conceivably some histamine present in the tissues might have been absorbed from the gut'. In the dog also, 'absorbed histamine must contribute to some extent to the origin of tissue histamine', according to Duncan and Waton (1969). In the guinea-pig, Schayer (1952) found only traces of ^{14}C-histamine in tissues 2 hours after its subcutaneous administration. Rats and mice given ^{14}C-histamine subcutaneously or intravenously retained measurable quantities of ^{14}C-histamine for at least 24 hours; large quantities were retained in the form of

MAGDALEN COLLEGE LIBRARY

imidazoleacetic acid and its riboside (Snyder, Axelrod & Bauer, 1964). A procedure employed in this study by Snyder *et al.*, in which butanol extractable ^{14}C is referred to as ^{14}C-histamine, is, perhaps, not fully reliable. This procedure was resorted to in Brodie's laboratory, where it was found that butanol-extractable radioactivity, when determined by isotope dilution, was no longer histamine, and as a result Brodie's laboratory retracted their earlier findings with butanol extracts (Brodie, Beaven, Erjavec & Johnson, 1966).

The mast cells are considered by some workers to make up the main bulk of histamine contained in the mammalian body. The lack of evidence for this belief is of little consequence for the following. All agree that in some tissues a major portion of the histamine is contained in mast cells. Furano and Green (1964) investigated the uptake of histamine by mast cells from the peritoneal fluid of the rat. *In vitro* $0.06–0.24 \times 10^6$ mast cells took up (from 1.47 µg of ^{14}C-histamine) histamine in a distribution ratio of $9:2$ between mast cell and medium. *In vivo* the mast cells likewise concentrated injected ^{14}C-histamine. In these experiments the rat lung also took up histamine. From their observations these authors discuss, amongst others, the following suggestions. 'Since histamine is absorbed from the gut of some species, notably the rat, exogenous histamine may contribute to the store in mast cells and therefore to the body stores of histamine.' 'The mast cell . . . may influence the amounts of amines to which the blood vessels and connective tissue are exposed.' Finally, the authors discuss the possibility that the mast cells may provide the tissues with subtle mechanisms to regulate local levels of histamine (and 5-HT). In our laboratory rat skin with intact mast cells did not detectably take up histamine from the blood, as will be described in Chapter 8.

In recent work, the most obvious object of the uptake has been overlooked and forgotten, namely the immediate removal from the blood of histamine which for some reason circulates in adverse concentration, e.g. anaphylaxis in some species. On this view, uptake is a preliminary and an adjunct to catabolism and excretion in the urine, and in this sense the process has a counterpart in the removal of catecholamines from the circulation by tissue uptake. Rose and Browne (1938) demonstrated that intravenously injected histamine disappears rapidly from the blood of the rat. In point of fact, three minutes after the injection, an estimated 90% of the injected histamine had been cleared away and taken up by the great number of tissues examined for uptake. This work was carried further in rats by Halpern, Neveu and Wilson (1959) who injected radioactive histamine, 500 μg/100 g, in place of 24 mg histamine per rat as employed by Rose and Browne. Even in the first minute after injection the blood concentration of radioactive histamine fell quickly. At 5 min after the injection the kidneys had taken up 146 μg/g, other organs examined also took up radioactive histamine, the heart, liver and lung in descending order. Halpern *et al.* concluded, as did Rose and Browne, that the rate of disappearance of histamine from the circulation is too fast to be explained wholly by its enzymic destruction and urinary excretion.

The uptake and retention for some time of large proportions of injected histamine by feline tissues has been demonstrated by Emmelin (1951) and by Lilja and Lindell (1961) (Table 11), as will be further discussed in Chapter 7. This uptake of exogenous histamine represents, as it were, an overspill which, even when retained intracellularly for some time seems to be devoid of the functions of endogenously formed histamine.

Physiological actions are exercised intracellularly by histamine in the cells which form the amine. There are exceptions from this, the most notable is what we refer to as the secretory device of the gastric mucosa which comprises the parietal cell and an adjacent histamine forming cell from which histamine appears bound to reach and impel the parietal cell. Even here there is no need of histamine uptake because the histamine forming component of the secretory device has a potentially very high HFC. Whenever additional histamine is required, this need is met by high rates of endogenous histamine formation. Indeed, the new approach to the physiology of histamine is based on this view.

Histamine can be absorbed from the gut, as shown by Mellanby (1916) who, in experiments undertaken in order to elucidate the condition of diarrhoea and vomiting of children, introduced 50 mg histamine into the intestinal lumen of cats and noted a subsequent fall in arterial blood pressure. Noteworthy in this publication is Mellanby's reflection regarding the disposal of the absorbed histamine: arguing by the analogy of other amines he assumed that imidazolylacetic acid constituted a step in the catabolism.

In anaesthetized dogs, Duncan and Waton (1968) introduced

[14]C-histamine into the lumen of various regions of the gastro-intestinal tract and recovered fractions of the introduced amount in the venous effluent. After the introduction of 50–5000 μg ^{14}C-histamine into a loop of jejunum, radioactive histamine was also detected in the general arterial blood. These authors suggest that 'the histamine of the food may be a source of tissue histamine and further that absorbed histamine may play a role in the gastric and intestinal phase of gastric secretion'. The present authors do not believe in a role for blood-borne histamine in gastric secretion, a view based on experiments described in Chapter 9.

Sjaastad and Nygaard (1967) have reported that in rats the introduction of 50 mg L-histidine into the stomach did not lead to an increase in the urinary output of free histamine, and that excessive increase in the flora of the small intestine *per se* is of minor consequence to the intestinal production of histamine.

Absorbtion of histamine from the gut has been demonstrated by several investigators. The amount absorbed is small, indeed trivial, in comparison with the amounts continuously formed endogenously in the various tissues. The problem arising with respect to absorbed histamine is, in Mellanby's phrase, how to dispose of it. The small amounts in question are rapidly cata-bolized and excreted.

In sum, uptake of exogenous histamine, where it takes place, provides a means designed to clear the blood very rapidly of other-wise adverse histamine, prior to its catabolism and excretion in the urine.

5

METHODS TO DETERMINE RATE OF
HISTAMINE FORMATION

In tissues of animals, bacteria and plants, several enzymes, under suitable conditions, are capable of forming histamine. On present knowledge histamine formation in these tissues takes place solely by the agency of decarboxylases which catalyse the conversion of L-histidine to histamine and carbon dioxide. The enzymes concerned are of two categories: histidine decarboxylase and enzyme(s) of the DOPA decarboxylase class. A physiological role has so far been recognized for the former enzyme only. The latter may well form histamine under natural conditions, although it has not, as yet, been possible to associate this latter mode of histamine formation with definite physiological events.

The tissue histidine decarboxylase activity can be measured *in vivo* and *in vitro*. Wherever possible, values obtained *in vitro* should, for safety, be matched with corresponding determinations *in vivo*. Further, to prove the identity of the enzyme, use should be made of available specific inhibitors of the two categories of enzyme concerned.

The methods to determine rate of histamine formation, i.e. histidine decarboxylase activity in individual tissues, or in the whole animal, are of these types:

(1) Non-isotopic determination of newly formed histamine;
(2) Isotopic: (a) whereby newly formed labelled histamine is measured; (b) determining $^{14}CO_2$ evolved from carboxyl labelled histidine. The general principles of these methods, and the methods for histidine decarboxylase assay have been comprehensively described by Schayer (1968a).

In the following the methods to determine histamine formation, as employed in our laboratory, will be set forth.

Non-isotopic whole-body histamine formation. In Chapter 3 it has been explained that the female rat, fed on a partly synthetic histamine-free diet, excretes free histamine in amounts which faithfully correspond to the rate of endogenous histamine formation and changes therein. The urinary histamine can be directly assayed on a piece of guinea-pig isolated ileum. Employing this simple method observations were made which brought histamine into the realm of normal physiology in that in the pregnant rat we found high rates of histamine excretion, the increased histamine originating from the foetuses. Further, during the formation of granulation tissue in skin wounds, the female rat excreted increased amounts of free histamine. These observations led us to the hypothesis that increased formation of histamine is associated with rapid growth, cell proliferation and the process of repair in certain tissues. Lastly, by determining the excretion of histamine we identified α-methyl histidine as a specific, non-toxic inhibitor of histamine formation.

It is held by Wetterqvist and White (1968) that the accuracy of the direct bioassay on urine containing small amounts of histamine, less than 40 μg/24 hr, is questionable. They have evidence to support the presence in urine of some material which renders the guinea-pig ileum less sensitive to histamine. In urine with a low concentration of histamine, where a larger volume of the sample is used for the assay, the dose-response curve of the gut was found slightly flatter than with the standard histamine solution. This situation was rectified by purifying the urine (rat) on a Dowex 50 column prior to bioassay, whereby slightly higher values for histamine were obtained than when the urine was assayed directly.

In man under standardized dietary conditions the urinary excretion of non-isotopic histamine, methylhistamine and methyl-imidazoleacetic acid has been investigated by Granerus (1968). The standard daily diet contained about 300 μg histamine (base). The values obtained are given in Table 4.

Among the results of Granerus it should be noted that administration of aminoguanidine which inhibits diamine oxidase was followed by a substantial increase in urinary methylhistamine and a small increase in 1,4-methylimidazoleacetic acid. In contrast to some other species, no sex difference in the pattern of urinary excretion of histamine and its metabolites seemed to be present. Smokers excreted larger amounts of 1,4-methylimidazoleacetic acid than non-smokers. The various techniques and methodo-

Table 4. Urinary excretion (μg/24 hr) of histamine (Hi), methylhistamine (MeHi) and methylimidazoleacetic acids under standardized dietary conditions in twenty healthy subjects

					During aminoguanidine treatment			
Subject	Hi	MeHi	1,4-MeImAA	1,5-MeImAA	Hi	MeHi	1,4-MeImAA	1,5-MeImAA
A	24	200	1800	7200	—	—	—	—
B	19	230	3600	2600	—	—	—	—
C	23	160	2800	5500	—	—	—	—
D	13	160	2900	n.m.	—	—	—	—
E	33	240	4500	800	—	—	—	—
F	28	60	2400	600	—	—	—	—
G	14	210	2600	1900	10	520	2900	1100
H	28	180	2300	500	—	—	—	—
I	31	190	2500	1800	—	—	—	—
J	14	110	1700	1100	15	470	2200	n.m.
K	6	110	2000	700	10	290	2800	n.m.
L	32	280	3800	1000	26	760	4300	1000
M	28	210	2300	n.m.	—	—	—	—
N	15	220	2400	n.m.	19	470	2900	800
O	14	180	2400	1200	17	580	2600	1600
P	46	130	2400	n.m.	—	—	—	—
Q	9	120	2000	1500	11	270	1900	600
R	37	350	2500	n.m.	—	—	—	—
S	15	200	2500	1400	14	360	2600	700
T	22	210	3700	5500	—	—	—	—

n.m. = not measurable.

(Granerus, 1968, *Scand. J. clin. Lab. Invest.* **104**, 59.)

logical considerations employed are fully described in Granerus' publication. Regarding nicotine it should be recorded that in the rat nicotine greatly enhances the rate of histamine formation in the gastric mucosa, but not in the lung or skin. The elevated mucosal HFC is paralleled by an increased urinary excretion of histamine (Svensson & Wetterqvist, 1968).

In this connexion it should be mentioned in passing that some alcoholic drinks are rich in histamine. Granerus, Svensson and Wetterqvist (1969) found values (expressed in μg base per 100 ml.) as follows: red Burgundy 1560, Chianti Vecchi 276, white Bordeaux 19, port 6, Moselle 4, beer 3–15, brandy and Scotch whisky 0. The authors reflect that presumably alcoholic intake should be controlled when studying histamine metabolism in man.

Non-isotopic in vitro *methods.* Following the demonstration of the presence of an enzyme, forming histamine from added histidine, in kidney slices of rabbits, by Werle (1936) and Holtz and Heise (1937), and now recognized as identical with DOPA decarboxylase

(Ganrot, Rosengren & Rosengren, 1961), Waton (1956) undertook an extensive study of what, at that time, he referred to as mammalian histidine decarboxylase. The procedure employed by Waton differs from that of the early investigators in that he added benzene to the incubation mixture. In the course of Waton's assay, mainly on minced rabbit kidneys, or tissue extract, he discovered that certain organic solvents, among them benzene, powerfully potentiated the decarboxylase activity. At the time of Waton's investigation it was not known that benzene potentiates DOPA decarboxylase activity and may inhibit histidine decarboxylase and, further, that on extracting tissues the latter enzyme loses in activity. Presumably, as a consequence of these circumstances Waton could not demonstrate histamine formation in tissues which in fact have been shown to contain histidine decarboxylase activity on a considerable scale. Waton's method has been used by several workers with the object of testing our hypothesis regarding an association between histamine formation and certain instances of rapid tissue growth.

We became interested in this method because of its failure to show histamine formation even in tissues exceedingly rich in histidine decarboxylase. Employing Waton's method, Kameswaran and West (1962) found that the mouse foetus 'is capable of forming only minute quantities of histamine from about the 15th day of gestation up to term (day 21) when both the liver and the kidney contribute much to the total activity'. Since in our laboratory, as will be described in a later chapter, foetal mouse histidine decarboxylase activity increased steadily during pregnancy to attain very high figures at term, and the results with other mouse tissues were largely at variance with ours, we decided to do some non-isotopic determinations in tissues with high HFC. Rosengren (1963) demonstrated, non-isotopically, substantial histamine formation in minced whole foetus. A further example of incompatible results: Mackay, Reid and Shephard (1961), non-isotopically, found no histidine decarboxylase in foetal liver and kidneys of the guinea-pig, whereas Kameswaran and West (1962), employing the same method, reported high rate of histamine formation in these tissues.

It should be recognized by now that the non-isotopic method, with benzene added, determines mainly DOPA decarboxylase, and that claims and conclusions relative to histidine decarboxylase based on this method are unjustified. If benzene is omitted, the non-isotopic method may be useful if caution is observed. The preformed histamine contained in all tissues constitutes a hazard, particularly in tissues with low HFC, since the newly formed histamine may represent a tiny amount as against the preformed histamine contained in the incubation vessel. In tissues with high HFC and low histamine content the benzene-free method may be used. The method could perhaps be applied in the search for inhibitors of histidine decarboxylase, provided precautions are taken to obviate any catabolism of newly formed histamine which could be mistaken for enzyme inhibition.

Isotopic methods employed in the author's laboratory

The approach by non-isotopic means opened the door for histamine to enter the realm of physiology. This approach did not permit exploration of the very inside, discovery of the place where histamine is essential to the machinery of the body. Such exploration will depend on methods of high sensitivity and specificity. These requirements are met by methods elaborated by Schayer. His methods emerged at a time when the lines of thinking in the field of histamine was at a juncture where these methods could be brought to proper use.

Schayer's methods have been fully described in his Handbook article (1966*a*) and in *Methods of Biochemical Analysis* (1968*a*). The *in vitro* method involves incubation of excised tissue with minute quantities of radioactive L-histidine, and determination of the ^{14}C-histamine formed. Schayer has added successive improvement on his original method, which still stands as the foundation of the various modifications.

Purification of ^{14}C-*histidine.* Since the isotopic method, employed *in vitro* or *in vivo*, determines the formation of ^{14}C-histamine from ^{14}C-histidine, the ^{14}C-histidine used should not be contaminated with ^{14}C-histamine. The ^{14}C-histidine obtained commercially contains small amounts of ^{14}C-histamine. This contamination makes necessary the use of blank samples for *in vitro* work and interferes with *in vivo* experiments. A rapid procedure for removing ^{14}C-histamine from ^{14}C-histidine has been devised by Kahlson, Rosengren and Thunberg (1963). This procedure reduces contamination to such an extent that the activity of blank samples is negligible. These authors (1963) described fully the methods used in our laboratory for determining histamine formation *in vitro* and *in vivo*.

Isotopic in vitro *determination*. Schayer's *in vitro* procedure as adapted for use in our laboratory has been described by Kahlson, Rosengren and Thunberg (1963). To measure HFC, excised tissues are minced with scissors, not homogenized or extracted, since destruction of cell structure is detrimental to histidine decarboxylase activity, a fact already noted by Schayer (1956a). The amount of substrate employed, 40 μg ^{14}C-histidine (incubated with about 0·2 g of tissue), is far too small for decarboxylases other than histidine decarboxylase to participate significantly in the formation of histamine as determined by this *in vitro* method. The participation of DOPA decarboxylase is *a priori* excluded. This is obvious from observations illustrated in Fig. 6. Rabbit kidney cortex extract, which contains DOPA decarboxylase, requires more than 2000 times higher concentrations of histidine to form histamine at rates equal to the histamine formation by histidine decarboxylase of foetal rat liver.

There are no standardized units for expressing histidine decarboxylase activity. In reports from our laboratory, tissue HFC is

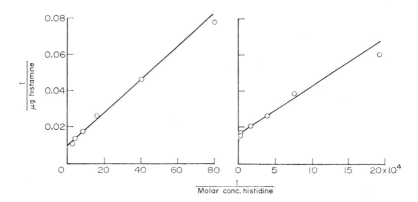

FIG. 6. Effect of substrate concentration on the rate of histamine formation in rabbit kidney cortex extract (to the left) and on the foetal rat liver extract (to the right). The reciprocal of amine formed per gram tissue in 3 hr is plotted against the reciprocal of histidine concentration (Ganrot *et al.*, 1961, *Experientia*, **17,** 263).

expressed in terms of the amount of histamine formed by 1 g of tissue in 3 hours. Schayer (1968*a*) observes that this 'practice might lead uncritical readers to make the unwarranted assumption that similar rates of histamine formation occur *in vivo*'. For this reason Schayer prefers to use a unit which shows relative rather than absolute rates of histamine formation.

Isotopic in vivo *determination.* As has been described previously the rate of endogenous histamine formation and changes therein can be followed by determining the urinary excretion of free histamine in female rats fed on a histamine-free diet. Similar results are obtained by an isotopic method devised by Schayer. His method measures the amount of ^{14}C-histamine excreted in the urine following the injection of a known amount of ^{14}C-histidine. This method determines whole-body HFC and can also be employed in measuring the *in vivo* HFC of individual tissues. In the latter case tissues are excised at chosen intervals after the injection of ^{14}C-histidine and their ^{14}C-histamine content is determined. Determining tissue ^{14}C-histamine at various intervals after injecting ^{14}C-histidine will, amongst others, give information on the life-time of newly formed histamine in individual tissues, as will be discussed later.

Isotopic ^{14}CO$_2$ *method of Kobayashi.* The isotopic techniques employed by Schayer and in our laboratory are laborious, time-consuming and require a considerable period of training. A simpler procedure for measuring histidine decarboxylase could be based on determining ^{14}CO$_2$ formation from ^{14}C-carboxyl labelled histidine, as described by Kobayashi (1963). Aures and Clark (1964) used this method in studies of inhibitors of histidine decarboxylase and reported, that with some exceptions, there is a correspondence between CO$_2$ evolved and histamine formed. Smith and Code (1967) made determinations on homogenized mouse kidney and felt confident with the ^{14}CO$_2$ method; this they acclaim as superior to other methods because it is believed to eliminate potential errors alleged to apply to methods based on the estimation of histamine formed by histidine decarboxylase. Maudslay, Radwan and West (1967) compared three different methods for the estimation of histidine decarboxylase activity *in vitro*: the isotopic dilution method of Schayer, as adapted for use in our laboratory, the $^{14}CO_2$ method of Kobayashi, and the non-isotopic method of Waton. They reported comparable results with the three methods. The authors concluded that the non-isotopic method can not be used for tissues with low enzyme activity, while Schayer's method is more sensitive but more complicated than the $^{14}CO_2$ method.

In our laboratory the formation of $^{14}CO_2$ from ^{14}C-carboxyl labelled L-histidine has been investigated (Grahn & Rosengren, 1968). The procedure employed was essentially a combination of the original method described by Kobayashi (1963) and its modification by Levine and Watts (1966). One object of the study was to compare Schayer's method (pipsyl method) with results obtained with the $^{14}CO_2$ method and the other was to see whether adrenaline would elevate mouse lung HFC *in vitro* in the same way as adrenaline was known to do *in vivo* in mice, but not in rats. Individual samples of minced lung were investigated concurrently by the pipsyl method and the $^{14}CO_2$ method. The results of these experiments are summarized in Table 5, which shows that the results obtained by the two methods differ.

Table 5. Seemingly formed histamine (ng/g) in minced mouse and rat lungs determined by two methods

| | Pipsyl method | | $^{14}CO_2$ method | |
	C	A	C	A
Mouse lung				
Expt. 1	100	160	170	130
2	120	200	210	190
3	70	110	120	120
Rat lung				
Expt. 1	160	160	130	110
2	160	150	140	120
3	150	150	190	180

Histidine concentrations were 0.8×10^{-4} M (pipsyl method) and 1.0×10^{-4} M ($^{14}CO_2$-method), and complete conversion to histamine would give 26·7 μg and 33·3 μg of histamine, respectively. In each experiment the figures represent the mean of two determinations on the same pooled tissue sample, corrected for non-enzymic formation of the recorded product except that attributable to adrenaline in the $^{14}CO_2$-method. C. Control; A, adrenaline, 5×10^{-4} M, added.

(Grahn & Rosengren, 1968, *Br. J. Pharmac.* **33**, 472.)

In samples of minced mouse lung, the amount of histamine seemingly formed was larger with the $^{14}CO_2$ method than with the pipsyl method. Discrepancies were also found with minced rat lung. Furthermore, adrenaline accelerated the formation of histamine in minced mouse lung as judged by the pipsyl method, whereas with the $^{14}CO_2$ method adrenaline seemed to depress histamine formation in minced mouse and rat lungs. Finally, in blanks containing adrenaline but no tissue, substantial amounts of $^{14}CO_2$ were evolved. Regarding results presented in Table 6 it should be noted that the ^{14}C-histidine concentration used for the $^{14}CO_2$ method was 20% higher than used for the pipsyl method. This difference is irrelevant in assessing the significance of the $^{14}CO_2$ method, and is due to the circumstance that in our laboratory determinations with the two methods were begun as independent studies.

The phenomenon of $^{14}CO_2$ formation in blanks not containing tissue was investigated further in the work under discussion. To reagent blanks not containing tissue, adrenaline was added in final concentrations of 10^{-5} to 10^{-3} M, and the yields of $^{14}CO_2$ and ^{14}C-ring labelled histamine formed from the two differently labelled histidines were determined with the $^{14}CO_2$ and pipsyl method, respectively. The formation of $^{14}CO_2$ from carboxyl labelled histidine was found to be directly proportional to the concentration of adrenaline. Substituting for adrenaline another reducing agent, sodium ascorbate, also increased the $^{14}CO_2$ formation from labelled histidine. By contrast, none of these compounds was found to cause the formation of significant amounts of ^{14}C-ring labelled histamine from ^{14}C-ring labelled histidine on incubation in idential conditions. These results are summarized in Table 6.

The finding that in reagent blanks histidine emits a substantial amount of CO_2 in the presence of adrenaline or sodium ascorbate calls for caution in the use of this method for assay of histamine formation. This non-enzymatic formation of $^{14}CO_2$ may occur also when tissue is present. The usefulness of the $^{14}CO_2$ method appears gravely limited by the finding that histidine apparently is transformed into some derivative which gives off CO_2 from the carboxyl group without concomitant formation of histamine. The reaction causing this phenomenon is unknown.

Table 6. Yields in terms of % $^{14}CO_2$ and ^{14}C-ring labelled histamine formed from the corresponding labelled histidines when reducing agents are present during incubation

		Yield (%) of labelled reaction product		
		Concentration of agent added (M)		
Agent added	Reaction product	10^{-5}	10^{-4}	10^{-3}
L-Adrenaline	$^{14}CO_2$	0·0223	0·130	0·74
	^{14}C-histamine	0·0085	0·0185	0·0069
Sodium	$^{14}CO_2$	0·051	0·48	3·03
L-ascorbate	^{14}C-histamine	0·0062	0·0108	0·0077

Incubations were carried out in Tris-HCl buffer and the gas phase was air in both types of experiment. All figures represent means of duplicates. A 100% yield would have been equal to 33·3 µg histamine base in the $^{14}CO_2$ method and 26·7 µg histamine base in the pipsyl method.

(Grahn & Rosengren, 1968, *Br. J. Pharmac.* **33**, 472.)

6

SOME CHARACTERISTICS OF MAMMALIAN HISTIDINE DECARBOXYLASE

Definition and terminology. Histidine decarboxylase (HD), L-histidine carboxy-lyase, E.C. 4.1.1.22 is the enzyme instrumental in the formation of endogenous histamine. The affinity of HD for histidine is very high, more than 2000 times higher than the affinity of the enzyme which has been extracted from kidneys of rabbits and guinea-pigs. This latter enzyme, discovered by Werle (1936) and by Holz and Heise (1937), is now recognized as identical with DOPA-decarboxylase, 3,4-dihydroxy-L-phenyl-alanine carboxy-lyase E.C. 4.1.1.26. An enzyme, named aromatic L-amino acid decarboxylase, first described by Lovenberg, Weissbach and Udenfriend (1962), can also decarboxylate histidine *in vitro*. It has a low affinity for histidine and is closely related to, or identical with, DOPA-decarboxylase. The enzyme(s) with low affinity for histidine is often denoted 'non-specific' histidine decarboxylase. The present authors are against referring to enzymes with a low affinity for histidine as histidine decarboxylase, even when given the prefix non-specific. These enzymes should, for the sake of clarity, be denoted DOPA-decarboxylase or its cognates. There is no evidence to show that the enzyme activity of this class contributes significantly to histamine formation in the normal *in vivo* conditions. This conclusion by Ganrot, Rosengren and Rosengren (1961) is supported by our observations on inhibition of histamine formation *in vivo*. Schayer and Sestokas (1965), on the other hand, found in the guinea-pig that the excretion of ^{14}C-histamine from injected ^{14}C-histidine is inhibited by α-methyl-DOPA and that the very minute amount of histamine excreted presumably was formed in the kidney.

Ubiquitousness of histidine decarboxylase. This enzyme has been found in all mammals and in every tissue examined by an adequate technique for the HD activity. Among the tissues indicated in Table 7 are some rich sources of the enzyme.

Table 7. Comparison of histidine decarboxylase activity (HFC) in some tissues of different species determined isotopically *in vitro* in our laboratory*

	Skin	Gastric mucosa	Maternal kidney	Placenta	Whole foetus
Rat	20	8000	10	10	3000
Mouse	30	1000	20,000	10	10,000
Hamster	20	1000	300	20,000	20
Guinea-pig	< 10	300	100	20	400
Cat	10	40		20	20
Man	10	40		< 10	

* The values are approximate and expressed as ng ^{14}C-histamine formed per gram of tissue.

(Kahlson & Rosengren, 1968, *Physiol. Rev.* **48**, 155.)

Tissue level of HD *activity.* In most tissues the activity of histidine decarboxylase is low and the amount of histamine continuously formed in tissues is correspondingly small. This should not be taken to imply that histamine formed endogenously at low rates is sub-threshold for action within the cell. In some species, e.g. rats, mice and hamsters, known for their low sensitivity to injected histamine, the whole-body formation of histamine, as judged by its urinary excretion, proceeds at considerable rates, even in the non-pregnant state. In the rat about 0·2–0·3 mg (in terms of the base) is formed in 24 hours, and in the mouse and hamster about 0·01–0·05 mg.

HD *synthesis can be rapidly accelerated.* Under the influence of various hormones, e.g. gastrin or catecholamines, the HD activity of target tissues rises rapidly (Figs. 26 and 47). A similar steep rise occurs with stimuli which provoke inflammation. The rise in HD activity engendered by the stimuli mentioned is prevented in the presence of inhibitors of protein synthesis such as puromycin and cyclohexamide (Schayer & Reilly, 1968; Snyder & Epps, 1968). This would suggest that elevated HD activity is the result of an increase in the amount of enzyme formed. In the terminology originally used by Schayer HD is an 'inducible enzyme' and the histamine thereby formed he referred to as 'induced histamine'. This terminology should not be taken literally since, by convention enzyme induction implies 'complete *de novo* synthesis of enzyme molecules which are new by their specific structure as well as by origin of their elements' (Jacob & Monod, 1961). More recently Schayer referred to histidine decarboxylase as having 'inducible activity'. In our laboratory, endogenous histamine associated with metabolic processes, e.g. growth and protein synthesis, is referred to as 'nascent histamine'. In instances where histamine emerges extracellularly in increased amounts, attention is primarily directed to the release of preformed histamine already contained in the tissues. The new terms serve the purpose of emphasizing the *de novo* formation of histamine at elevated HD activity, as distinct from the release of preformed histamine.

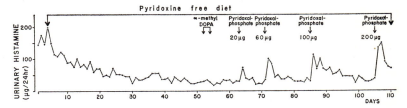

FIG. 7. Effect of pyridoxal-phosphate on the urinary excretion of free histamine in a female rat on a pyridoxine-free diet and receiving aminoguanidine. The α-methyl-DOPA (4 × 100 mg/ kg daily) was injected subcutaneously for 2 days (Kahlson, Rosengren & Thunberg, 1963, *J. Physiol.* **169**, 467).

Co-enzyme of histidine decarboxylase. Evidence for pyridoxal phosphate as a co-enzyme was provided by Rothschild and Schayer (1958a, 1959) who in rat peritoneal mast cells found a large loss of HD activity after dialysis. The activity could be restored by the addition of pyridoxal phosphate. Similar results were obtained by Ono and Hagen (1959) with HD extracted from mouse mastocytoma tissue. This co-enzyme function has also been shown *in vivo* in our laboratory. Female rats are fed a histamine-free diet and on omitting pyridoxal phosphate, the formation of histamine is greatly reduced, as judged by its rate of excretion. The urinary excretion falls within the first days, and then falls gradually to about 50% of normal within 2 weeks, and to about 20% in 6–7 weeks (Fig. 7). The rate and degree of the fall depend on how free from pyridoxine the casein in the animals' diet is. The casein used in our study contained <0.65 mg/kg; the rat received <3 µg pyridoxine daily, as compared with 300 µg in the adequate diet. The original rate of formation is rapidly restored when pyridoxal phosphate is added to the diet or injected subcutaneously. An initial rebound above this rate is commonly observed (Fig. 16). Similar results were obtained with the isotopic *in vivo* method (Kahlson & Rosengren, 1959; Kahlson, Rosengren & Thunberg, 1963).

From the experiment outlined in Fig. 7 it would appear that any compound that interacts with pyridoxal phosphate and deprives HD of this co-enzyme will inhibit histamine formation.

We conducted experiments *in vitro* in order to clarify to what extent the results thus obtained would agree with those obtained *in vivo*. Extracts of rat gastric mucosa were dialysed for various periods in a cellophane tube against flowing tap water in a rocking dialyser. Loss of activity occurred which could not be fully restored by the addition of pyridoxal phosphate. The nature of the non-restorable loss in activity could not be established; it was not restored by the addition of Fe^{3+}-ions, reported to be essential in addition to pyridoxal phosphate, for the activity of histidine decarboxylase from *Lactobacillus* (Guirard & Snell, 1954). To determine the role of pyridoxal phosphate alone the dialysed extracts were divided into two parts; pyridoxal phosphate was added to one part, the other half served as controls, and the enzyme activity of the two was compared in order to establish the difference in activity due solely to this co-factor. The optimum concentration for added pyridoxal phosphate was 10^{-5} M. The loss in activity, occurring during 200 hr of dialysis, and then still due solely to the removal of pyridoxal phosphate, amounted to about 75%. In the experiments by Kahlson, Rosengren and Thunberg (1963) HD activity *in vitro* was determined isotopically and whole-body HFC was determined non-isotopically using urinary histamine as an index. Results obtained with the two methods were in accord.

Inhibitors of histidine decarboxylase. α-methylhistidine has been recognized as a specific inhibitor of HD and is now widely used to distinguish HD from DOPA decarboxylase. This subject will be discussed in Chapter 8.

Miscellaneous characteristics. Following the demonstration, in 1958 and 1960, of a conspicuously high HD activity in the rat foetus, particularly high in the liver, a considerable amount of work has been done with the object of characterizing this enzyme. Except for the experiments with pyridoxal phosphate and inhibitors, which were carried out *in vivo*, studies on the characteristics of HD were undertaken *in vitro*. The results have been summarized extensively by Burkhalter (1962), Schayer (1966*a*), Niedzielski and Maśliński (1968), and by others. In terms of physiology, i.e. the situation existing in intact tissues, data obtained *in vitro*, particularly with tissue extracts, may not be representative. This applies, amongst others, to *in vitro* determination of pH optimum and HD-substrate relationship. In most cases neither the pH inside the cell nor its content of free histidine can be ascertained. Regarding the characteristics, behaviour and physiological significance of histidine decarboxylase, the present authors unreservedly side with the following general statement by Professor Chain in a lecture: 'Biochemistry done on complete organs or tissues is likely to yield results nearer to the reality of living animals than can be got from enzyme preparations'. He proceeds to say: 'Altogether I have wondered for some time how significant the results obtained from studies with cell-free enzyme systems, and the various general metabolic schemes which have been suggested, really are for the living organism' (Chain, 1965).

7

CATABOLISM OF HISTAMINE

THE principal routes involved in the degradation of histamine were clarified in the 1950's, mainly by the work of Schayer, who has summarized these studies in reviews of catabolism of histamine *in vivo* (Schayer, 1959; 1966*b*). Fig. 8, representing Schayer's view, shows the various ^{14}C-labelled catabolites excreted in the urine after the injection of a small amount of ^{14}C-histamine. It is assumed, although difficult to prove, that small amounts of exogenous histamine are catabolized in the same way as endogenous histamine. Otherwise their functional status is perhaps not the same in that endogenous histamine seems to exert functions which cannot be reproduced by exogeneous histamine.

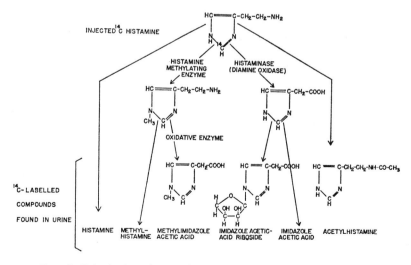

FIG. 8. Principal pathways for the catabolism of injected ^{14}C-histamine.

Enzymes capable of inactivating histamine are widespread in mammals and even in plants. In mammals the principal means of catabolism are methylation and oxidative deamination. In germ-free rats we found evidence of acetylation of endogenous histamine, as mentioned in Chapter 3. The occurrence of this pharmacologically inactive conjugate of histamine has subsequently not been thoroughly investigated and most of the pertinent experiments were made decades ago (reviewed by Tabor, 1954). Histamine interacts with co-enzymes I and II to produce histamine nucleotides. These interactions and products, the significance of which at present is not understood, are being studied mainly by Alivisatos whose review (1966) of this topic should be consulted.

METHYLATION

In male rats, and in mice, guinea-pigs, cats, dogs, and humans of both sexes, methylhistamine, 1-methyl-4-(β-aminoethyl)-imidazole, is a major metabolite of histamine (literature until 1959 reviewed by Schayer, 1959). Part of the methylhistamine formed is excreted in the urine, while another excreted part has been oxidized to 1-methylimidazole-4-acetic acid. Recently a predominant methylation has been shown also in the hamster (Rosengren, 1965). The methyltransferase(s) operating in these catabolic steps and a means of *in vivo* inhibition of the methylation of histamine have only recently been established. This topic will be further discussed in the chapter on histamine and the brain (Chapter 12). Reilly and Schayer (1970) have shown that methylhistamine inhibits histamine methylation *in vivo* in mice as evidenced by the observation that pretreatment with non-isotopic methylhistamine increased the levels of ^{14}C-histamine in the liver following injection of ^{14}C-histidine.

In the species in which methylation predominates, i.e. the majority of investigated species, the rate of whole-body histamine formation and changes therein has been determined by measuring urinary excretion of radioactive methylhistamine after injection of radioactive histidine and by concurrent determination of the fraction excreted as unchanged free histamine. Non-isotopic methods to determine methylated compounds have recently been developed: Fram and Green (1965) and White (1966) determined methylhistamine, Granerus and Magnusson (1965) and Tham

(1966) methylimidazole acetic acid, part of the work being done by gas chromatography. Since the non-isotopic methods explore the catabolism of endogenous histamine, they provide a means to see whether the catabolism of endogenous and exogenous histamine proceed by identical pathways. These methods are likely to promote studies in man in which methylated compounds are the principal catabolites excreted and found in tissues. Results obtained with these methods will be discussed.

HISTAMINASE (DIAMINE OXIDASE) (DIAMINE: O_2OXIDOREDUCTASE (DIAMINATING) E.C. 1.4.3.6)

The study of histaminase is a story of itself. The controversy over its identity and its properties as compared with those of diamine oxidase will not be discussed here, since the question is irrelevant to the subjects presented in this book (for discussion on this topic see Kapeller-Adler, 1965; Zeller, 1965). Histaminase or diamine oxidase refers here to the enzyme(s) metabolizing histamine by oxidative deamination, which can be strongly inhibited *in vitro* and *in vivo* by aminoguanidine.

There is a revival of interest in histaminase as seen in the review article by Buffoni (1966) and current clinical work related to this enzyme. A comprehensive monograph on amine oxidases has been produced by Kapeller-Adler (1970).

Histaminase in pregnancy. The presence of a histamine-inactivating factor in the human placenta (Danforth & Gorham, 1937) and in the blood of pregnant women (Marcou, Athanasiu-Vergu, Chiricéanu, Cosma, Gingold & Parhon, 1938) represents the first recorded instance of a major change in histamine metabolism in normal physiology. This discovery instigated a search for the biological significance of the histamine-inactivating agent. Hopes were entertained of throwing light on the significance of tissue histamine via studies of this factor in the same way that studies on choline esterase have yielded valuable information about the functions of acetylcholine. However, the significance of an elevated histaminase activity in human pregnancy proved difficult to reveal. The sensitivity of pregnant and non-pregnant women to injected histamine did not differ discernably in the following tests: dizziness and headache (Janowitz & Grossman, 1949), size of the flare of the triple response (Wicksell, 1949*a*), eosinopenic response (Kullander, 1952) and gastric secretory response (Clark & Tankel, 1954). These and other observations led some workers to believe that the elevated plasma histaminase activity found *in vitro* during pregnancy was not associated with an increased inactivation *in vivo*. Later experiments, to be described presently, showed that the above mentioned tests, as then employed, were unsuitable for the purpose of assessing differences in the blood histaminase concentration. Törnqvist (1968) infused histamine intravenously in pregnant and non-pregnant women; as tests he employed the increase in pulse rate and the increase in forearm blood flow. With these tests the pregnant women were significantly less sensitive to histamine when intravenously infused.

Pregnant women dispose of injected histamine more efficiently. The means by which this is achieved are not fully understood. Lindberg (1963*a*) showed that intravenous infusion of ^{14}C-histamine resulted in a lower blood concentration of ^{14}C-histamine when a woman was pregnant, and Lindberg and Törnqvist (1966) noted that administration of aminoguanidine to pregnant women raised the concentration of infused ^{14}C-histamine in the blood.

In the human foetal circulation, and in the human foetal tissues proper, the predominant route of histamine degradation is that of enzymic methylation. On injecting ^{14}C-histamine into one artery of the umbilical cord (the foetus remaining intact in the uterus) to pass through the placenta, less than 0.2% of the injected histamine appeared as free histamine in the urine, and the pattern of excreted metabolites indicated methylation as the principal metabolic route. Further, human foetal tissues incubated with ^{14}C-histidine formed ^{14}C-histamine, and methylated added ^{14}C-histamine (Lindberg, Lindell & Westling, 1963*a*, *b*).

Investigations of the urinary pattern of radioactive histamine catabolites have given results which are difficult to interpret, and, perhaps, do not reflect plainly physiological events. The work by Lindberg *et al.* (1963*a*, *b*) was preceded by an investigation by Nilsson, Lindell, Schayer and Westling (1959) of the metabolism of ^{14}C-labelled histamine in pregnant and non-pregnant women. The main results of this investigation are seen in Table 8. The increased plasma histaminase activity of the pregnant women did not significantly change the capacity to oxidize injected histamine

Table 8. Metabolism of injected ^{14}C-histamine in three non-pregnant and four pregnant women. One of the non-pregnant women, subject A, was given two injections, the first one intravenously and the second one, 6 months later, subcutaneously.

	Non-pregnant women				Pregnant women			
Subject, age in years	A, 25	A, 25	B, 26	C, 34	D, 45	E, 24	F, 26	G, 32
Month of pregnancy	—	—	—	—	VI	V	VI	V
Amount of ^{14}C-histamine injected (μg)	10	10	8·4	10	10	10	11·2	10
Route of injection	intravenous	subcutaneous	subcutaneous	subcutaneous	subcutaneous	subcutaneous	subcutaneous	subcutaneous
Fraction of injected ^{14}C found in 12 hr urine (%)	91	99	79	81	62	81	79	91
Values are expressed as per cent of amount of ^{14}C excreted in 12 hours								
Histamine	3	3	2	2	3	1	2	2
Imidazoleacetic acid total and (free)	34 (—)	34 (22)	25 (23)	37 (—)	36 (12)	53 (44)	26 (10)	34 (—)
Methylhistamine	4	6	4	2	2	3	1	2
Methylimidazoleacetic acid	53	51	39	30	38	34	19	23
Acetylhistamine	—*	1	—	1	—	—	2	—
Total	94	95	70	72	79	91	50	61

* — = not measured.

(Nilsson, et al., 1959, Clin. Sci. **18**, 313.)

as judged by the excreted metabolites. The limited value of study-ing the pattern of radioactive urinary metabolites after injecting ^{14}C-histamine is further emphasized by the observation that the pattern of urinary metabolites of ^{14}C-histamine in the pregnant human females included in Table 8 is mainly the same as the pattern found by Schayer and Cooper (1956) in human males. It should be emphasized that the cited results pertain to the catabolism of exogenous ^{14}C-histamine. The catabolism of endogenous hista-mine, employing non-isotopic methods, remains a topic for further study.

In the rat an enormous increase in the histaminolytic power of the uterus and placenta occurs during pregnancy, as described by Roberts and Robson (1953). It is a remarkable fact that in the rat even when not subjected to inhibition of histaminase by amino-guanidine, large amounts of the histamine produced by the foetuses do traverse the seemingly powerful histamine-destroying placental and uterine barriers. Other instances of circumvention of effective functional interrelation between histamine and histaminase occurring in the same tissue will presently be discussed.

Distribution of histaminase in cats. This species was chosen by Haeger and Kahlson (1952a) for study because the cat is very sensi-tive to injected histamine and has a pattern of histaminase dis-tribution similar to man, as judged by the few available deter-minations on the latter. Further, the cat was considered suitable in investigating the hormonal control of tissue histaminase activity. In the cat, histaminase activity is present mainly in the small intestine and kidney (Table 9). Histaminase activity was deter-mined by the method of Wicksell (1949b) which measures the amount of added histamine inactivated in a definite period of time.

In the intestine the bulk of histaminase is found in the glandular layers, and only little is contained in the muscular layers. In the gastric mucosa no histaminase was found.

Histamine is contained in all regions of the alimentary tract. In the cat the small intestine contains about as much histamine as the gastric mucosa. In the small intestine the ratio between histamine in the glandular and muscular layers varies between 1:1 and 5:1 per unit weight of tissue. There is no correlation between the intestinal contents of histamine and histaminase, that is to

Table 9. Histaminase activity expressed in μg histamine base destroyed by
1·0 g of tissue in 1 hour (fed cats)

Cat no.	Stomach	Small intestine	Colon	Rectum	Kidney
1	< 0·25	120	30	7	80
2	< 0·25	96	28	10	96
3	< 0·25	76	16	6	96
4	< 0·25	52	30	6	80
5	< 0·25	80	24	24	92
			Colon + rectum		
6	—	120	11		—
7	—	96	26		—
8	< 0·25	—	4		80
9	—	96	26		—
10	—	72	10		—
11	—	120	—		—
12	< 0·25	96	2·5		72
13	< 0·25	96	9		78
14	< 0·25	120	15		120
15	—	120	12		84
16	< 0·25	56	2		88
17	< 0·20	98	6		96
18	< 0·20	120	10		120
19	< 0·20	144	—		84
20	< 0·25	168	—		156
21	< 0·20	48	1·1		—
22	< 0·20	150	10		150
23	—	96	28		96
24	—	56	2		98
25	< 0·25	89	2·5		—
26	—	120	—		94
		100·2 ±6·15(25)			97·9 ±5·31(19)

(Haeger *et al.*, 1952*a*, *Acta Physiol. scand.* **25**, 230.)

say, high figures for histaminase do not correspond to low figures
for histamine and vice versa. Histamine in a conjugated form could
not be demonstrated in the intestinal mucosa.

Histaminolytic activity of lymph. The failure to demonstrate the
occurrence of histamine in the venous effluent during reactive
hyperaemia and muscular tetanus led us to search for histamine in

the lymph. In the course of these studies it was noted that the lymph is protected against the accumulation of histamine much more efficiently than is the blood, and that the lymph, unlike plasma, is very active in destroying histamine. These observations were made in non-pregnant dogs (Carlsten, Kahlson & Wicksell, 1949a). Lymph was collected from a cannula inserted into the thoracic duct. Normal lymph did not contain histamine in detectable amount, probably because its histaminolytic activity was high. In fifteen dogs the mean value in the lymph was 1·5 μg histamine (base) inactivated by 1 ml. in 1 hour, as against a mean value in the plasma of <0·07 μg/ml./hr. In the lymph nodes the concentration of histaminolytic activity was about the same, or somewhat higher, than in the lymph. On intravenous infusion of 25–400 μg histamine per minute during 15–90 minutes the histamine concentration of plasma rose to very high levels, but there was no rise in lymph histamine. On infusing histamine, 1·5–100 μg/min, for 10 to 20 min into the central stump of a mesenterial or femoral lymphatic trunk, only a small fraction of the infused histamine, or none, was recovered in lymph collected from the thoracic duct.

The origin of the histaminase present in the lymph from the thoracic duct was investigated by Carlsten (1950a). First, he demonstrated that the thoracic duct lymph from the cat, dog and rabbit of either sex, in contrast to plasma, contains histaminase in considerable concentrations. Next, he cannulated lymph ducts from various regions and organs and found that the lymph emerging from the intestine and the kidneys was singularly rich in histaminase, whereas in cervical lymph the content was low. Carlsten reflects on the question why and for what purpose histaminase from these organs is flushed into the lymph and is steadily poured via the thoracic lymph duct into the blood stream, where it can hardly be detected at all.

Histaminase and hormones. The finding that in the rat, adrenalectomy is followed by a fall in the histaminase content of the lung, and that this change can be restored to normal by an adrenocortical extract (Karady, Rose & Browne, 1940) incited Carlsten (1950b) to investigate the effect of adrenalectomy on lymph and plasma histaminase in cats. In ten out of thirteen cats of both sexes investigated adrenalectomy was followed by a striking increase in the histaminase content of the thoracic duct lymph, as shown in

Fig. 9. This increase was maximal within 2 to 3 hours after adrenalectomy and persisted for about 24 hours, whereafter the values were normal at 48 and 72 hours after adrenalectomy owing to depletion of the histaminase stores. In the plasma no histaminase activity could be detected either before or after adrenalectomy. In plasma obtained after adrenalectomy no agent could be detected endowed with the property of inhibiting *in vitro* the histaminolytic activity of the lymph continuously discharged into the general circulation.

Carlsten and Wood (1951) showed that in cats a continuous infusion of adrenaline did not reverse the increased histaminase activity of thoracic duct lymph after adrenalectomy, whereas infusion of an adrenocortical extract was effective in reversing the change.

It remained to investigate whether the histaminase entering the lymph after adrenalectomy was in fact derived from the main stores of the enzyme, the intestine and the kidneys in cats. Experimental evidence showed that adrenalectomy or hypophysectomy caused a profound reduction in the content of histaminase in the intestinal mucosa and the kidney. Within one hour after adrenalectomy, and 2 to 4 hours after hypophysectomy, large amounts of histaminase emerged into the lymph channels while the histaminase content of the main stores was reduced to a small fraction of the original (Haeger, Jacobsohn & Kahlson, 1952). Here, again, the large amounts of lymph histaminase continuously discharged into the blood disappeared in an unknown way. In this study it was also investigated whether changes in the concentration of histamine would occur on the disappearance of the bulk of tissue histaminase. Hypophysectomy or the removal of the adrenal glands did not appreciably alter the concentration of histamine in the mucosa of the small intestine and stomach as compared with that of normal cats. However, the total content of histamine present in the mucosa was greatly reduced on account of the ensuing atrophy of the mucosa, a phenomenon which will briefly be described in an Appendix to this chapter.

The rapid disappearance of the histaminase discharged by the lymph channels into the blood remains a mystery. In cats the viscera contain about 90% of the histaminase present in the whole body. After adrenalectomy more than 75% of this histaminase disappears. This loss could be due to the absence of a prosthetic

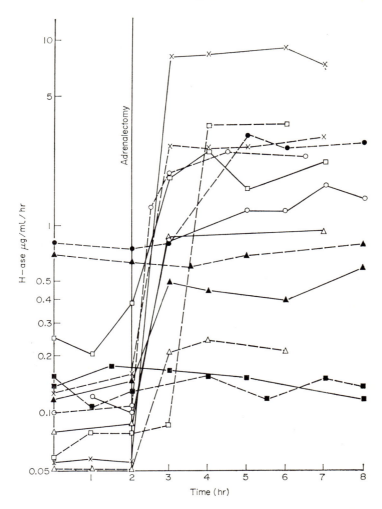

Fig. 9. Histaminolytic activity of thoracic duct lymph in thirteen cats under chloralose, before and after adrenalectomy. Semi-logarithmic curve (Carlsten, 1950*b*, *Acta physiol. scand.* **20,** Suppl. 70, 33.)

group or of some other activator of the enzyme, a possibility which cannot be tested because these factors have not been identified. An alternative to this has been tested, namely the shifting over of histaminase to other regions of the body. With this in view tissues of young cats (400–600 g) were divided into two parts and homogenized, the one part comprising the thoracic, abdominal and pelvic viscera, the other part the rest of the body except the head and the skin. After adrenalectomy histaminase disappeared from the viscera, but in the 'rest of the body' the histaminase content was not significantly altered and in the skin none could be detected (Haeger & Kahlson, 1952*b*). The figures in Table 10 stand for 1 μg histamine (base) inactivated by 1 gram in 1 hour.

Table 10. Histaminase content in normal and adrenalectomized cats

Exp. no.*	Normal			Exp. No.*	Adrenalectomized		
	Viscera	Rest of the body	Whole body		Viscera	Rest of the body	Whole body
1a	630	68	698	1b	17	144	161
				1c	39	124	165
2a	1488	149	1637	2b	67	86	153
3a	960	100	1060	3b	55	15	70
4a	1790	84	1874	4b	375	69	444
				5	40	28	68

* The cats used for every individual number of experiment are members of the same litter.

(Haeger *et al.*, 1952*b*, *Acta Physiol. scand.* **25**, 255.)

After adrenalectomy excess histaminase appeared in the lymph within 1 hour; the delay following hypophysectomy was 2–3 hours (Fig. 10) (Haeger, Jacobsohn & Kahlson, 1953*a*). Otherwise the derangement in the metabolism of histaminase was fundamentally the same after adrenalectomy and hypophysectomy. It appears that the adrenal cortex depends upon stimulation from the hypophysis for the production and release of agents controlling the metabolism of histaminase. The experiments of Fig. 10 indicate that the body's own supply of ACTH is rapidly exhausted after hypophysectomy, and that an infusion of ACTH restores the ability of the stores to retain their histaminase. After adrenalectomy, cortisone but not ACTH or DOCA, restores histaminase to the normal situation.

The pituitary–adrenal control of histaminase exists in other species also. In the guinea-pig the pattern of distribution is different from the cat. In the guinea-pig the tissues contain histaminase in the following descending order: liver, small intestine, kidney, spleen. The female kidney contains less than the male one

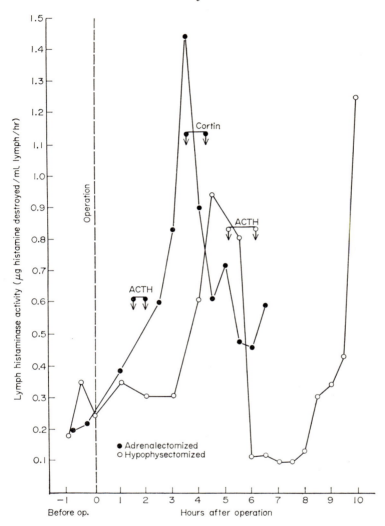

FIG. 10. The adrenalectomized cat received an infusion of 1·3 mg. ACTH acid peptide (Dr. Li) over 30 min, and of adreno-cortical extract (2 ml., Cortin, Organon) over 45 min. The hypophysectomized cat was given an infusion of 1 mg ACTH peptide during 60 min (Haeger *et al.*, 1953a, *Acta physiol. scand.* **30,** Suppl. 111, 170).

(Lindell & Westling, 1953). Adrenalectomy in the guinea-pig was followed by a fall in histaminase activity in the liver (50% fall) and kidneys, but not in the small intestine. This latter tissue was in this respect resistant to the consequences of adrenalectomy in the same way as was the 'rest of the body' in the cat. As in the cat, cortisone, but not DOCA, prevented the fall in histaminase activity (Kahlson, Lindell & Westling, 1953).

In sum: The retention of histaminase in distinct depots, and presumably also its formation, is controlled by the pituitary–adrenocortical system.

Changes in histaminase with age. In the small intestine of the feline foetus the concentration of histaminase is approximately the same as in the mother. Among foetuses of the same mother the range of variation is very small. After birth the values scatter within a wider range. In the kidney of the feline foetus the histaminase concentration is only about 10% of that of the mother. After birth the concentration increases as demonstrated in Fig. 11. The adult level is reached at about the eighth month (Haeger, Kahlson & Westling, 1953). In the human foetus the distribution of histaminase is similar to that in the feline foetus (this laboratory, unpublished). It should be mentioned in this connexion that the plasma amine oxidases in pigs and goats are absent at birth and start to develop during the 20 to 30 days after birth (Blaschko & Bonney, 1962).

Histaminase and heparin. It has been believed that, apart from pregnancy, no other condition exists in human beings in which the concentration of histaminase in the plasma is greatly increased. Recently it was discovered by Tryding (1965) that intravenous administration of heparin in man brings about an increase of the histaminase activity of the blood plasma to levels which otherwise are seen only during the later stage of pregnancy. In Tryding's experiments heparin had no effect on the histaminase activity of plasma or whole blood *in vitro*. This effect of heparin has been studied in some detail by Hansson, Holmberg, Tibbling, Tryding, Westling and Wetterqvist (1966). In blood plasma the normal values were 0·005 to 0·025 units (u.) per litre per minute, rising during pregnancy to final values between 1 and 7 u./l. per minute. (A unit is defined as the enzyme activity which converts 1 μM of

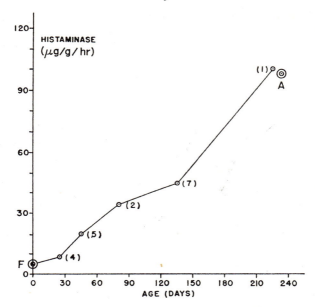

Fig. 11. Histaminase concentration in the kidneys of kittens 25–220 days of age. F =mean value for 20 foetuses; A =mean value for 19 adult cats, 98 units (Haeger and Kahlson, 1952a). Figures in parentheses denote number of kittens (Haeger, Kahlson & Westling, 1953, *Acta physiol. scand.* **30**, Suppl. 111, 177).

the substrate ^{14}C-putrescine per minute). Lymph, collected from a cannula inserted in connexion with thoracic surgery into the thoracic duct near the venous angle, contained 0·5–0·9 u./l. per minute. The lymph/plasma ratio in man agreed with the corresponding histaminase values found in the dog by Carlsten *et al.* (1949a). On injecting 10,000 units of heparin intravenously in a 37-year-old woman the histaminase level in the thoracic lymph rose to 140 u./l. per minute. The increase in plasma was less than a hundredth of that in the lymph. In man the intestinal mucosa, the kidneys and the liver are the main deposits of histaminase (Zeller, 1942).

In the guinea-pig in which the liver is the main store of histaminase (Lindell & Westling, 1953), injections of heparin deplete this store, and excess histaminase appears in the plasma (Hahn, Schmutzler, Seseke, Giertz & Bernauer, 1966). The depleted store is rapidly restored by resynthesis (Schmutzler, Bethge & Moritz, 1967). In the experiments of Hahn *et al.*, surprisingly, injection of protaminsulphate also depleted the store, but, curiously, when heparin and protamin were injected together, they neutralized each other, and no increase in plasma histaminase occurred.

In the rabbit, i.v. injection of heparin elicited a dose-dependent increase in plasma histaminase, noticeable within $\frac{1}{2}$ min, reaching maximal values within about 1 hr and then consecutively declining to basal levels. This phenomenon occurred also in new-born animals (Hansson & Thysell, 1970).

Figure 12, from Dahlbäck, Hansson, Tibbling and Tryding (1968), shows the time-course of the increase in histaminase in blood and lymph following injection of heparin in a male patient. Before heparin injection the ductus thoracicus lymph, which is drained into the blood stream at the rate of several litres in 24 hours, had a histaminase concentration about thirty times that of plasma.

The increase in histaminase in the blood during human pregnancy is believed to be derived, at least partly, from the decidual cells of the placenta (Swanberg, 1950). In pregnant women the effect of heparin on histaminase in blood plasma and lymph has been investigated by Hansson, Tryding and Tornqvist (1969). The heparin-induced additional increase in histaminase activity in the plasma in pregnant women did not differ significantly in magnitude from that in non-pregnant women. This increase in histaminase activity of the plasma disappeared with a half-time of 70 to 80 minutes, a situation similar to that in non-pregnant women. The histaminase content of thoracic duct lymph appeared to remain unchanged during pregnancy, despite the conspicuous increase in plasma. There was no evidence that the histaminase content of the placenta is influenced by heparin injection.

The physiological significance of histaminase

Histaminase is instrumental in the inactivation of histamine *in vivo*. This can be shown easily in the female rat which excretes free histamine. Owing to the limited life-time of intracellular histamine, a steady stream of extracellular histamine exists, part of which is excreted. If the histaminase activity is inhibited by administering aminoguanidine, the urinary excretion of histamine increases about three-fold.

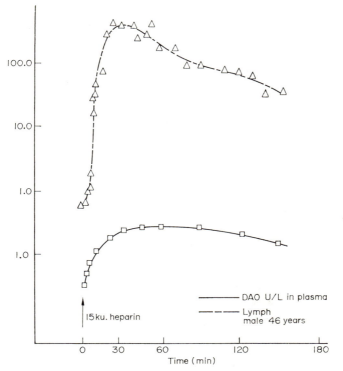

Fig. 12. Diamine oxidase (DAO) activity in blood plasma (squares, continuous line) and thoracic duct lymph (triangles, broken line) in patient G.G. before and after intravenous injection of 15 kilo-units of heparin (at time zero) (Dahlbäck, Hansson, Tibbling & Tryding, 1968, *Scand. J. clin. Lab. Invest.* **21**, 17).

The action of histaminase on extracellular histamine can be demonstrated on organs *in vitro*. In 1947 Grandjean reported that thiamin (vitamin B_1) potentiated the contraction of the guinea-pig's ileum produced by histamine added to the bath. Thiamin is an inhibitor of histaminase (Zeller *et al.*, 1939) and the guinea-pig's small intestine contains histaminase (Lindell & Westling, 1953; Arunlakshana, Mongar & Schild, 1954). Semicarbazide inhibits histaminase (Zeller, 1942) and potentiates the contraction produced by histamine on the guinea-pig's ileum preparation (Mongar & Schild, 1951).

In the cat, the small intestine and the kidneys are rich in histaminase. Emmelin (1951) infused increasingly large amounts of histamine intravenously for several hours in cats and determined the ensuing elevation of the histamine content of arterial plasma. Large amounts of the injected histamine were removed by the viscera, as seen by the additional steep elevation of arterial plasma histamine concentration which occurred on evisceration. Tissues other than the viscera removed some of the infused histamine. The kidney took up and retained histamine for some time. On infusing histamine, 40 μg/kg per minute, into a renal artery for 5 minutes (a total of about 500 μg histamine) and killing the cat one minute after discontinuing the infusion, the kidney contained 40 μg histamine/g, whereas the other kidney which had been removed before the onset of histamine infusion contained only 0·6 μg/g.

An experiment similar to Emmelin's was carried out on eviscerated cats by Arunlakshana, Mongar and Schild (1954) who recorded the fall in blood pressure after histamine injections with and without administration of the histaminase inhibitor B_1 pyrimidine. In only four of the ten cats tested did the enzyme inhibitor potentiate the fall in blood pressure after histamine injections; these results again testify that in the cat histaminase resides mainly in the viscera which had been removed.

Lindell and Westling (1956) demonstrated that in cats after the administration of histaminase inhibitors, aminoguanidine being the most potent among four inhibitors tested, the depressor response to histamine injected into the renal artery was consistently potentiated. If, however, in the experiments of Lindell and Westling histamine was injected into the femoral artery or vein after the cat had been given aminoguanidine, no potentiation of the depressor response was observed. Besides the kidney and intestine, other organs metabolize, and even take up exogenous histamine. This was shown by Lilja and Lindell (1961) who investigated the metabolism of injected ^{14}C-histamine in heart-lung-liver preparations and heart-lung preparations from cats. Their object was to assess the capacity of the liver, the heart and the lungs to take up and metabolize injected histamine and to retain the amine and the metabolites. Into these two kinds of preparation 32 μg ^{14}C-histamine (corresponding to about 65,000 c.p.m. under the conditions of assay) was injected. In the heart-lung-liver preparation most of the injected histamine disappeared from the blood within 0·5 to 2 min after the injection. Tissue samples investigated 30 min after the injection contained large amounts of ^{14}C-histamine and its metabolites (Table 11). Inactivation by histaminase played a minor part in these organs, as seen from the low values for imidazoleacetic acid. Methylation predominated and this is in accord with Lindahl (1958, 1960) who studied *in vitro* the methylating enzyme in the liver, and also in agreement with Brown, Tomchick and Axelrod (1959) who demonstrated the presence of this enzyme in other tissues such as the heart and lung. In the heart-lung preparation, ^{14}C-histamine injected into the inferior vena cava disappeared more slowly from the blood than in the preparation in which the liver was included. Determinations made 30 min after the injection showed that the major part of the injected histamine was retained in the blood, heart and lungs. Methylhistamine was the predominant metabolite, especially in the heart.

Table 11. The distribution of ^{14}C-histamine and some of its metabolites in the blood and tissues of two heart-lung-liver preparations

Experiment A	Blood counts/min	Heart (20 g) counts/min	Lungs (10 g) counts/min	Liver (86 g) counts/min	Total counts/min
Histamine	500	280	110	700	1580
Methylhistamine	2000	200	100	1400	3700
Methylimidazoleacetic acid	30,000	3600	1900	30,800	66,300
Imidazoleacetic acid	750	0	30	420	1200
	33,250	4080	2130	33,320	72,780
Experiment B	Blood counts/min	Heart (15 g) counts/min	Lungs (20 g) counts/min	Liver (80 g) counts/min	Total counts/min
Histamine	0	160	40	160	360
Methylhistamine	0	390	0	0	390
Methylimidazoleacetic acid	24,000	1950	2300	20,000	48,250
Imidazoleacetic acid	150	0	20	240	410
	24,150	2500	2360	20,400	49,410

Samples taken 30 min after the injection of 32 μg ^{14}C-histamine into the portal vein. (Lilja & Lindell, 1961, *Br. J. Pharmac.* **16**, 203.)

The guinea-pig, the favourite species for studies of anaphylaxis, has not been extensively studied for its histamine metabolism. Schayer obtained evidence to indicate that in this species oxidation and methylation occurred to approximately the same extent and that exogenous histamine was not stored (for references see

Schayer, 1966*b*). To our knowledge the distribution of histaminase and its inhibition *in vivo* in this species has been investigated only by Lindell and Westling (1953, 1954). As inhibitor, they employed stilbamidine which had been shown by Blaschko *et al.* (1944, 1951) to inhibit histaminase strongly. Histaminase resides predominantly in the liver and injection of stilbamidine substantially lowered the histaminase activity as seen in Table 12. Injection of stilbamidine in doses which inhibited the histaminase, potentiated the broncho-constriction and increase in urinary bladder tone caused by histamine given intravenously.

Table 12. Mean values for histaminase activity in organs from the guinea-pigs used. Figures within parentheses denote number of animals

| | Histaminase activity (μg/g/hr) | | | |
| | | Dose of stilbamidine (mg/kg body weight) | | |
Tissue	No stilbamidine	1–5	5–10	10–16·5
Liver	15·1 (10)	10·3 (6)	6·7 (6)	3·3 (5)
Lung	0·4 (10)	0·3 (4)	0·2 (5)	0·1 (5)
Kidney	2·0 (10)	1·1 (6)	0·6 (6)	0·3 (5)
Bladder	6·8 (10)	5·1 (4)	4·1 (5)	3·5 (5)

(Lindell & Westling, 1953, *Acta physiol. scand.* **30**, 202.)

In the report on the discovery of conjugated histamine in the urine by Anrep and his colleagues it is stated that dogs do not excrete histamine in the free form, not even when fed on meat (Anrep, Ayadi, Barsoum, Smith & Talaat, 1944). In the dog only the kidneys have been thoroughly examined for histaminase activity. Lindell (1957) demonstrated that inhibitors of histaminase potentiated the depressor responses to histamine infused into the renal artery; no potentiation occurred when histamine was in-jected into the femoral artery or by vein. This latter fact pertains also to the cat in which, as already described, potentiation was absent because of the effective uptake and subsequent catabolism in other organs. In Lindell's experiments the kidney retained a considerable capacity to remove injected histamine from the blood even after the administration of histaminase inhibitors, indicative of the presence in the kidney of histamine removing means other than histaminase. The situation was clarified by Lindell and Schayer (1958) who infused 1 or 4 μg ^{14}C-histamine per minute into the renal artery and collected urine for analysis from tubes inserted into the ureters. Strong evidence was obtained that the kidney oxidizes the greater portion of the histamine to imidazoleacetic acid, a result which is consistent with the *in vitro* results by other workers, among them Waton (1956), who reported a very high histaminase activity of the dog kidney. The kidney receiving ^{14}C-histamine, excreted, besides a major portion of imidazoleacetic acid, a considerable portion of free ^{14}C-histamine and smaller amounts of methylated compounds.

In the rat, in which histaminase is the principal histamine catabolizing enzyme, the histaminolytic power of the uterus (Roberts & Robson, 1953), and placenta (Ahlmark, 1944; Anrep, Barsoum & Ibrahim, 1947) is high during pregnancy. Even in the plasma of pregnant rats an elevated histaminase activity has been found (Kobayashi, 1964). Nonetheless, large amounts of histamine produced by the foetuses indeed traverse the seemingly powerful histamine-destroying placental and uterine barriers, to be excreted in the urine. The function of this histaminase build-up was at times considered strategically designed to protect the foetuses and the uterus against any influx of excessive amounts of histamine whatever its origin. The histamine-destroying action of histaminase can be inhibited without interference with the course of pregnancy and without observable harmful consequences to mother and litter (Kahlson, Rosengren & Westling, 1958).

In the cat the foetus forms histamine at a much lower rate than in the rat. At no stage of pregnancy could Carlsten (1950*c*) detect histaminase activity in placental extracts or plasma in the cat.

In the pregnant woman, as in the non-pregnant, histaminase is a considerable histamine catabolizing enzyme. On intravenous infusion of 0·1 μg ^{14}C-histamine (base) per kg body weight per minute, Lindberg (1963*a*) made the following observations on the content of ^{14}C-histamine and its metabolites in the blood. The ^{14}C-histamine concentration of systemic arterial blood was about 50% lower when the woman was pregnant. As mentioned earlier, analysis of blood from the umbilical vein showed that only small amounts of ^{14}C-histamine had passed unmetabolized through the placental barrier. In systemic blood and in the uterine vein imidazoleacetic acid was the predominant metabolite in pregnancy, whereas in non-pregnancy methylhistamine and methylimidazole-acetic acid were in excess. *In vitro* determinations by Lindberg (1963*b*) showed that in tissues of placenta, and even in pregnant myometrium, oxidative deamination to imidazoleacetic acid by histaminase, was the major route which could be blocked by aminoguanidine, but in non-pregnant myometrium methylation to methylhistamine and methylimidazoleacetic acid were dominant.

In order to study *in vivo* routes of inactivation of histamine reaching the placenta via the foetal circulation, Lindberg, Lindell and Westling (1963*a*) injected ^{14}C-histamine into an umbilical artery in pregnant women who underwent legal abortion. Contrary to the results obtained *in vitro*, the greatest amount of ^{14}C excreted was not in the form of imidazoleacetic acid, but was in the form of methylated histamine derivatives. As a possible explanation of their result, Lindberg *et al.* refer to Lindahl (1960) who demonstrated the presence of imidazole-N-methyl transferase in extracts of human placenta, and who on the basis of her findings believes that in the human placenta *in vivo*, methylation may be the major route of inactivation.

Nonetheless, it has been demonstrated that the histaminase activity of pregnant women which is so prominent when determined *in vitro*, does exist *in vivo* also. It is disturbing and unexplained why this fact escapes demonstration on injecting ^{14}C-histamine and determining the metabolites excreted, as in the experiment of Table 8.

The means by which histamine is inactivated in the human is not known in its entirety, except for the demonstration that this inactivation takes place at exceedingly high rates. Of pertinent experiments, the results obtained by Adam, Card, Riddell, Roberts and Strong (1954), who infused histamine (3·6–72 ng/kg per minute) of which only about 1% of the dose given appeared in the urine, should be mentioned. Regarding the loci of histamine inactivation, information has been obtained in man of both sexes by intravenous infusion of ^{14}C-histamine until a steady blood concentration was obtained and hence determining the ^{14}C-concentration in blood from the pulmonary artery, the hepatic, renal, internal jugular, iliac and axillary veins: the largest extraction of injected ^{14}C-histamine occurred in the portal and renal circulation (Arnoldsson, Helander, Helander, Lindell, Lindholm, Olsson, Roos, Svanborg, Söderholm & Westling, 1962).

Histaminase unerringly contrives to capture the mind of investigators, at times even resorting to solitary approaches. Having noted that the rhesus monkey exhibits higher plasma histaminase activity than any other species tested and stressing the importance of subhuman primates as models of man in biological research, Peters and Gordon (1968) determined the histaminase activity of plasma in sixteen species of subhuman primates, including gorilla and chimpanzee of both sexes (non-pregnant). In the majority of the species the values were very high, in a few only the enzyme activities were as low as in man. These authors ascribe important biological meaning to plasma histaminase and suggest that for investigations of drugs or toxic agents expected to influence the physiological or pharmacological role of histamine in man, New World subhuman primates may be better test animals than Old World ones.

On the preceding pages only the inactivation of injected histamine has been discussed. Intracellular histamine becomes extracellular after having led its lifetime and is thereafter likely to be metabolized by the same routes as small amounts of injected histamine. It should, however, be emphasized that this presumption has not been proved. Pending the elaboration of non-isotopic methods of determining the principal histamine catabolites in the urine and tissues, uncertainty exists over this issue. Indeed, all available evidence indicates that intracellular histamine can be sustained at a steady level independent of gross changes in histaminase activity. Experiments have been described above in which adrenalectomy or hypophysectomy in cats was followed by an up to 90% loss of histamine destroying activity of the intestinal mucosa without a concomitant alteration of the histamine concentration of that tissue. In this instance the intracellular tissue histamine concentration is not influenced by the histamine destroying power of the tissue, the concentration of histamine did not increase although the amine is continuously newly formed. The fact that the histamine concentration remains constant can be explained by the operation of the 'regulatory mechanism' discussed in Chapter 3 and by the disposition of the amine and the catabolizing enzyme in separate compartments of the cell as in this case first shown by Waton (1956).

In sum: What then, if any, is the physiological significance of histaminase? One function appears well established: histaminase, in conjunction with methylating enzymes, inactivates the histamine which enters the blood stream and hence is carried through the body. In physiological conditions, to the best of present knowledge, histamine never occurs in the blood in concentrations apt to produce harmful effects. There is no evidence that the pertinent histaminase build-up is essential to the normal course of pregnancy and parturition in man and animals. It would thus appear that the body machinery is endowed with this enzyme activity, at least partly, for other purposes than that of disposing of histamine.

It has been demonstrated by Smith (1967) that placental histaminase is capable of oxidizing a number of amines, several aliphatic diamines even faster than it oxidizes histamine. This would imply a more fundamental significance of histaminase in cell metabolism. An alternative function of histaminase at this level finds support in the fact that even some seedlings produce histaminase (Werle & Pechmann, 1949, reviewed by Werle, 1964). Pea seedlings contain diamine oxidase the substrate and inhibitor specificities of which resemble placental histaminase (Zeller, 1963; Hill & Mann, 1964).

To the physiologists histaminase remains a challenge because its level of activity is conspicuously increased, normally in pregnancy, artificially by heparin, and by some unknown way disappears from the blood when entering there via the lymph, is inhibited *in vivo* by aminoguanidine, and also because its disappearance and reappearance (resynthesis) are so overwhelmingly under hormonal control. To the prospective adherer to what in some quarters is referred to as molecular biology, the challenge will lie in the fact that the physiological significance of histaminase is unknown.

APPENDIX: ATROPHY OF INTESTINAL MUCOSA FOLLOWING HYPOPHYSECTOMY OR ADRENALECTOMY

In a previous section it was described that the removal of the hypophysis or the adrenal glands in cats deprives the intestinal mucosa of the ability to retain its store of histamine destroying activity. The loss was immense in magnitude, 90% or even more. The histaminase drained away via the lymph channels into the blood stream from where it rapidly disappeared.

In the course of this work in which the mucosa had to be

collected and weighed, we were bound to note that the gastric and intestinal mucosa had undergone grave atrophic changes. Persistent students of the physiology of histamine have often, quite unexpectedly, been faced with something new, outside the centre of their chosen field of endeavour. The observations made in our laboratory are clearly seen in Fig. 13 (facing p. 104). It was concluded that the maintenance of the structural integrity of the gastrointestinal mucosa is dependent on the pituitary gland and that the hypophyseal control is, at least partly, mediated by the adrenal glands (Haeger, Jacobsohn & Kahlson, 1953*b*). This suggestion has been confirmed by Crean (1968) who demonstrated that hypophysectomy in the rat markedly reduced the weight of the stomach and volume of the fundic mucosa, and the total parietal cell population.

8

INHIBITION OF HISTAMINE FORMATION
AND WHAT IT REVEALED

ENDEAVOURS to establish the physiological significance of a certain class of agents, such as acetylcholine and catecholamines, have so far been guided by three modes of approach: the study of physiological effects caused by release of the particular agent, and by blocking its release, or the administration of antagonists which specifically obviate changes normally effected by the agent. None of these approaches has been rewarding in attempts to elucidate physiological roles for histamine. A fourth approach remains, inherently decisive, if it were feasible: to inhibit the very formation of the agent.

Until recently, no means were known to inhibit histamine formation *in vivo* and could not even be attempted because methods for *in vivo* HFC determinations were not available. In independent studies and at about the same time, two *in vivo* methods were elaborated: Schayer's isotopic method in which ^{14}C-histamine formed from injected ^{14}C-histidine was determined in urine or tissues, and the non-isotopic method which depends on the recognition that the urinary excretion of free histamine, in the female rat fed a special diet, reflects the rate of endogenous histamine formation and changes therein. These methods have been described in Chapter 5.

In the experiments on inhibition now to be discussed, issues of experimental technique and basic principles are involved. Prior to the *in vivo* experiment, the animals should be given several days to become accustomed to the metabolism cage, since urinary histamine values usually decline for a period before a steady base line is reached. The state of feeding is also important, in that if rats are not fed, urinary histamine is reduced, owing to lowering of gastric mucosal HFC, as described in Chapter 9 on gastric secretion. Both of these effects could be mistaken for enzyme inhibition. More intricate is a basic issue involved. Inhibition of what kind of histamine formation should be studied: inhibition of the overall endogenous histamine formation, or inhibition of histamine formation from injected histidine. *A priori* it would appear that the former approach is closer to physiology than the other, which, as will be described, gives inconsistent results.

The experiments and conclusions to be discussed in this chapter have been published in two reports by Kahlson, Rosengren and Thunberg (1963) and by Johnston and Kahlson (1967). A few of our unpublished observations will also be included.

Cortisone and prednisolone

At the beginning of our work on inhibition of histamine formation *in vivo*, Schayer, Smiley and Davies (1954) had reported that injection of cortisone inhibits histamine formation in the skin but not in the stomach of rats. Injection of prednisolone was recorded greatly to reduce the histidine decarboxylase activity of particle-free extracts of excised rat lung (Schayer, 1956*b*). In our laboratory neither three successive daily injections of 1 mg cortisone nor prednisolone in six daily doses of 5 mg significantly altered the formation of whole-body histamine, as reflected in its urinary excretion to which histamine from the lung contributes only a minor part. These results should not be taken as at variance with Schayer's who found glucocorticoids to reduce HFC in some rat tissues, but to increase it in stomach, so overall, there might be a balance.

Pyridoxine deficiency and semicarbazide

Semicarbazide, a carbonyl reagent, was shown by Werle and Heitzer (1938) to inhibit guinea-pig kidney decarboxylase *in vitro*, an enzyme now recognized as of the DOPA decarboxylase class. Since we strongly desired an animal with lowered HFC, we started our inhibitory *in vivo* studies employing semicarbazide. Daily subcutaneous injections of this agent in doses gradually increasing from 50 to 500 mg/kg caused an initial slight and inconsistent reduction in histamine formation, even when this agent was administered for as long as 6 weeks (Fig. 14). On discontinuing semicarbazide, the rate of histamine formation, as reflected in its urinary excretion was rapidly restored and occasionally exceeded the original rate, sometimes by multiples. This 'overshoot', a state which we have found useful in experiments on wound healing and collagen formation, to be described in Chapter 19, may perhaps indicate that the inhibition of histidine decarboxylase is over-ridden by a concurrent formation of new molecules of the enzyme which become fully active on discontinuing semicarbazide to which the cofactor is believed to be attached. Pyridoxal phosphate as a cofactor of histidine decarboxylase has been discussed in Chapter 6. In our experiments (1963) extracts of gastric mucosa were dialysed for 200 hr which deprived the histidine decarboxylase of about 75% of its activity by the removal of pyridoxal phosphate. Desoxypyridoxine is referred to as a pyridoxine antagonist (see Umbreit & Waddell, 1949). We injected thirteen daily doses of 150 mg/kg in rats and found no significant effect on the whole-body HFC. This observation, again, indicates a firm attachment of the coenzyme in intact tissue cells. The pharmacological properties of pyridoxal phosphate have been reviewed by Holtz and Palm (1964) and its biochemical significance by Dixon and Webb (1964).

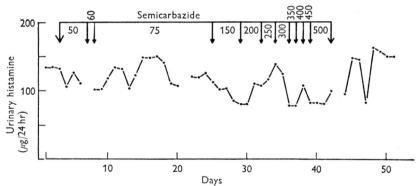

Fig. 14. Urinary histamine in a female rat on aminoguanidine. Semicarbazide, in the doses (mg) indicated, given daily for 40 days (Kahlson, Rosengren & Thunberg, 1963, *J. Physiol.* **169,** 467).

Notwithstanding our failure to demonstrate inhibition of HFC by semicarbazide *in vivo*, we thought it justified to study its effect *in vitro*. We employed our standard procedure and determined ^{14}C-histamine formed from ^{14}C-histidine by minced rat gastric mucosa in 3 hr. Semicarbazide revealed strong inhibitory potency.

Figure 15 shows the effect on histamine formation by varying the concentration of semicarbazide, 50% inhibition being produced by 5×10^{-5} M, and 99·8% by 10^{-2} M semicarbazide. Next, we determined the degree of inhibition of histamine formation in the minced mucosa, ear and lung with the object of seeing whether histidine decarboxylase of these tissues would be more resistant to inhibition. As shown in Table 13 semicarbazide produced similar inhibition in the three tissues. *In vivo*, by contrast, a dose of 112 mg/kg, which, if evenly distributed would correspond to 10^{-3} M, failed to decrease the urinary excretion of histamine in rats fed on a diet adequate in pyridoxal phosphate.

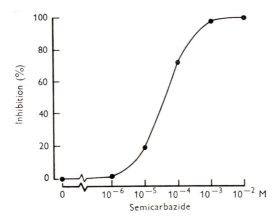

FIG. 15. Inhibition *in vitro* by semicarbazide of ^{14}C-histamine formation in minced rat gastric mucosa (Kahlson, Rosengren & Thunberg, 1963, *J. Physiol.* **169**, 467).

Table 13. Inhibition *in vitro* of histamine formation in three different tissues by semicarbazide contained in the incubation mixture in the concentrations stated. The values refer to counts/min/gram tissue and, in parentheses, % inhibition

	Concentration of semicarbazide		
	0	5×10^{-5} M	10^{-2} M
Ear	150	57 (62)	2 (98·7)
Lung	1400	660 (53)	3 (99·8)
Gastric mucosa	76,000	33,000 (57)	90 (99·9)

(Kahlson, Rosengren & Thunberg, 1963, *J. Physiol.* **169**, 467.)

The 'escape' from inhibition *in vivo*, especially the overshoot in histamine formation on discontinuing semicarbazide, we interpreted, by mere guess-work, as the result of the inhibition being overtaken by a concurrent formation of new molecules of histidine decarboxylase. These new molecules, we conjectured, would not exhibit full potency in the absence of the cofactor. Thus, for complete lack of other means, we decided to inject semicarbazide in pyridoxine deficient rats. The results vindicated our expectation. In the type of experiment exemplified in Fig. 16, the pyridoxine-deficient diet and injections of semicarbazide were instituted simultaneously from the very first day. As a result whole-body histamine formation fell rapidly to about 20% of normal within 2 days, even with doses of semicarbazide as moderate as 2×50 mg/kg subcutaneously per day. With larger amounts injected over longer periods the animals exhibited signs of frustration, excessive irritability, seizures, loss of appetite and body weight. This procedure to inhibit histamine formation *in vivo* is not practicable in mice because semicarbazide is more toxic in this species than in rats (Kahlson, Rosengren & Steinhardt, 1963*a*).

It will be seen in what follows that so far no other means

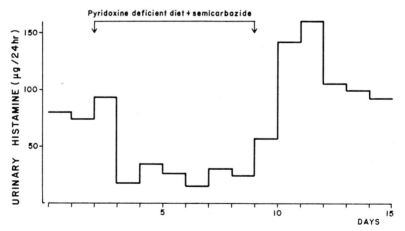

FIG. 16. Urinary histamine in a female rat on aminoguanidine. Between the arrows pyridoxine-deficient diet and injections of semicarbazide subcutaneously in daily doses of 2×50 mg/kg. Note the overshoot of histamine excretion on discontinuing inhibition (Kahlson, Rosengren & Thunberg, 1963, *J. Physiol.* **169**, 467).

employed equals semicarbazide superimposed on the pyridoxine-deficient diet in its power to inhibit histamine formation *in vivo* in the rat. This result has been confirmed by several workers, among them Johnson (1969), who in comparison with other means, found this one the most potent and reliable method of inhibiting histidine decarboxylase *in vivo*. With this procedure other pyridoxine-dependent decarboxylases also will become less active, to a various and not established degree, a circumstance not requiring further explanation. Disregarding the lack of specificity, inhibition by this procedure gave insight into relevant circumstances which hitherto were unknown, and some of which will now be discussed:

1. *Relation between tissue histamine content and* HFC. At the time of our report on inhibition, published in 1963, it had been disclosed by Schayer, and in our laboratory, that a consistent relation between histamine content and HFC of tissues did not exist. This relation we studied more closely. Rats were subjected to inhibition of histamine formation for 13 to 199 days by pyridoxal-deficiency either alone or combined with semicarbazide. The resulting inhibition of whole-body histamine formation was of the order of 80–90%. The histamine content of various tissues was determined at the 13th–199th day of inhibition. Before the onset of inhibition one ear and a sample of abdominal skin were removed to serve as controls. Additional normal values included in Table 14 (brackets) are taken from Gustafsson, Kahlson and Rosengren (1957). From this table it appears that in the abdominal skin, lung and tongue the histamine content is not significantly altered following strong inhibition of histamine formation for 13 and 22 days. On very prolonged inhibition, for 143 and 199 days, the histamine content was lowered in the abdominal skin and the tongue, but not in the lung. The skin and tongue are tissues rich in mast cells (Riley, 1959). In the ear, also rich in mast cells, the histamine content was lowered at the 13th day of inhibition, and then fell continuously during the course of prolonged inhibition. In the small intestine the histamine level persisted during 13 and 22 days of inhibition, and was lowered only after very long periods of inhibition. These observations on the small intestine are noteworthy on account of the reported absence there of mast cells and because of its low HFC.

MAGDALEN COLLEGE LIBRARY

Table 14. Histamine content in rat tissues, µg (base)/g

Group	Deficient diet (days)	Ear	Abdom. skin	Tongue	Lung	Gastric mucosa	Small intest.
N		[23·6]	24·2	[36·2]	[5·0]	31·2	[23·9]
A	143	7·7	7·3	18·7	4·8	8·8	2·8
		(27·0)	(20·0)				
A	143	—	—	23·1	4·6	2·8	3·2
A	199	3·3	7·7	20·9	8·4	18·2	3·2
		(31·2)	(23·0)				
B	13	10·4	23·7	42·4	12·9	3·6	55·5
		(16·3)	(22·6)				
B	13	9·3	14·5	38·8	7·1	—	23·7
		(14·4)	(16·9)				
B	22	—	—	41·0	5·7	1·5	16·4
B	22	—	—	36·9	7·4	2·1	11·5
B	22	—	—	37·2	5·9	2·1	19·8
B	75	7·9	21·9	36·6	6·6	0·9	10·8
		(19·3)	(28·4)				
B	75	14·5	21·5	37·0	11·5	1·1	6·3
		(26·0)	(24·1)				
C		14·7	—	39·0	7·7	—	8·5
		(29·6)					
C		12·2	—	—	—	—	—
		(25·0)					

N, normal rats; mean of 8 determinations on abdominal skin and gastric mucosa, figures in brackets from Gustafsson *et al.* (1957); A, pyridoxine-free diet; B, pyridoxine-free diet and semicarbazide (2 × 75 mg/kg daily) for the last 5 days; C, adequate diet for 4 days, preceded by pyridoxine-free diet for 31 days and semicarbazide (2 × 75 mg/kg daily) during the last 5 days of deficiency. Figures in parentheses are determination of an ear and a small piece of abdominal skin removed before institution of inhibition. — stands for not examined or lost sample.
(Kahlson, Rosengren & Thunberg, 1963. *J. Physiol.* **169**, 467.)

In the gastric mucosa, a predominantly non-mast-cell tissue with a high HFC, the situation is entirely different. Here the histamine content fell to low values, particularly low following maximum inhibition. The observations presented in Table 15 show that even one day of inhibition caused a precipitous fall in mucosal histamine content and HFC, with a tendency to further fall in the content with more prolonged inhibition.

In our inhibition study (1963) we investigated six tissues with the object to correlate more directly their histamine content and

Table 15. Histamine content and histamine-forming capacity (HFC) of gastric mucosa in rats submitted to concomitant pyridoxine deficiency and semicarbazide, 2×75 mg/kg daily for, respectively, 1, 2, 3 and 4 days

	Content (μg base/g)	HFC (counts/min per gram)
Normal	31·2*	81,000†
1 day's inhibition	7·5	8000
	8·9	
2 days' inhibition	6·2	700
	6·5	
3 days' inhibition	5·1	1900
	8·2	
4 days' inhibition	4·3	2500
	7·6	
1 day's adequate diet, preceded by 4 days' inhibition	32·1	76,000
	34·0	

    * Mean of 8 specimens.   † Mean of 2 specimens.
      (Kahlson, Rosengren & Thunberg, 1963, *J. Physiol*, **169**, 467.)

HFC at a degree of inhibition of whole-body histamine formation of about 80%, as judged from the urinary excretion. In addition, further evidence appeared desirable for verifying whether with the mode of inhibition employed, the HFC of the six tissues would be equally or differentially depressed. The results are given in Table 16. As in the *in vitro* experiments represented in Table 13 there was no obvious indication of a differential inhibition.

The figures given in Table 16 reveal an embarrassing circumstance: when the six tissues were excised and examined for their HFC at the stage of approximately 80% whole-body *in vivo* inhibition, a significantly higher degree of inhibition prevailed in the determinations *in vitro* (Residual HFC, Table 16). This apparent discrepancy emphasizes the pitfall encountered in studies of inhibition of histidine decarboxylase, and calls for a warning against judging the degree of *in vivo* inhibition of histamine formation by determinations only on excised tissues, as this may lead to overrating the inhibitory potency.

2. *Feed-back relation.* An initial lowering of tissue histamine content presumably elevates tissue HFC by a feed-back relation between histamine content and histidine decarboxylase activity, a repression which operates *in vivo* but not *in vitro*. Accordingly, the

Table 16. Comparison of HFC and histamine content, determined on the same pooled samples of various tissues. In two experiments, each on four adult female rats from the same litter, the tissues were pooled (*a*) from two control rats and (*b*) from two rats 'inhibited' by pyridoxine deficiency for 13 days and by semicarbazide (2×75 mg/kg, subcutaneously) for the last 5 days

Tissue	HFC (counts/min per gram)			Histamine content (μg base/g)		
	Control	Inhibited	Residual HFC (%)	Control	Inhibited	Inhibited as % of control
Ear	260	19	7	24	33	137
	260	11	4	32	28	88
Abdominal skin	110	7	6	22	29	132
	110	6	5	20	29	145
Tongue	190	14	7	50	45	90
	200	10	5	58	54	91
Lung	1850	90	5	8	12	150
	2300	200	9	6	11	183
Gastric mucosa	17,900	380	2	37	6	16
	5600	540	10	41	6	15
Small intestine	8	0·2	3	28	—	—
	10	0·7	7	—	20	—

(Kahlson, Rosengren & Thunberg, 1963, *J. Physiol.* **169**, 467.)

degree of inhibition manifested *in vivo* would represent the balance resulting from inhibited and concurrently newly formed enzyme molecules. In point of fact, in this inhibition study (1963) we found indications of a feed-back relation of this kind, which later could be more firmly established as operating in the gastric mucosa.

In rats on an adequate diet a 'long-acting' histamine preparation (the poorly soluble histamine dipicrate) was injected subcutaneously, 5 mg/rat on 3 successive days. The tissues were thus flooded with histamine (the repressing end-product) to an extent which increased the urinary histamine to a maximum of about 5·5 times the normal value. Subcutaneous injections of ^{14}C-histidine were given to rats in the normal state and at the height of flooding with histamine. Urinary ^{14}C-histamine was determined on 3 successive days after the ^{14}C-histidine injection. The amounts of ^{14}C-histamine excreted were lower than expected, indicating that flooding the tissues with histamine exerted a repressing influence on histidine decarboxylase activity, thus suggesting that the amine contributes to the control of its own rate of biosynthesis. This suggestion was fully substantiated in our later study of the feed-back relation existing in the gastric mucosa, where it was shown that this relation operates in both directions, i.e. lowered histamine content brings about elevated histidine decarboxylase activity (Chapter 9).

3. *Lifetime, mode of binding*. Pertinent information on these topics has been obtained by two different approaches, the one by Schayer and the other as employed in our inhibition study (1963). Schayer (1952) and Schayer, Smiley and Davies (1954) injected ^{14}C-histidine into guinea-pigs and rats and determined the amount of ^{14}C-histamine present in excised tissues at various time intervals. In rat abdominal skin the rate of turnover of histamine was about 2 μg/g in 24 hr, corresponding to a lifetime of about 15 days. These figures compare well with observations by Feldberg and Talesnik (1953) that restoration of skin histamine content after depletion by a liberator is a slow process, taking about 2–3 weeks. In the guinea-pig, in the combined pool of histamine in the lungs, intestines, and kidneys, Schayer (1952) found a half-life of the order of 50 days.

As already mentioned, we examined various tissues for histamine content at 1 to 199 days of strong inhibition of histamine formation. In tissues rich in mast cells, such as the abdominal skin and tongue, inhibition of histamine formation to a small fraction of the normal for weeks was not followed by a significant fall in histamine content (Table 14). By contrast, in the gastric mucosa, a predominantly non-mast-cell tissue, the histamine content fell to low levels within 24 hr (Table 15). That is to say, in mast cells a histamine molecule once formed is firmly held and has a long intracellular lifetime, whereas in the gastric mucosa the reverse is true. In mast-cell tissues the histamine content depends on the lifetime and number of binding sites and less on the HFC, whereas in a non-mast-cell tissue, such as the gastric mucosa, the actual histamine content represents the balance between the rate of formation and rate of removal of histamine.

In the rat lung, believed to be poor in mast cells, the HFC was found to be relatively high, but the histamine content was low. This situation appears to indicate that the major portion of newly formed histamine leaves the lung rapidly. A smaller portion appears to be retained in structures that possess a type of binding providing for a long lifetime. This proposition implies the existence of non-mast-cell tissue with histamine in two distinct pools, as judged by the extreme differences in lifetime.

The small intestine, in which mast cells are rare and HFC conspicuously low, was found to retain its high histamine content after 3 weeks of strong inhibition of histamine formation. Only in the course of a very long period of inhibition did the content fall to low levels. Here, too, a mode of intracellular binding different from the mast-cell type must be sought.

An approach to this problem similar to that of Schayer and his colleagues was employed by Bjurö, Westling and Wetterqvist (1964*a*) with the view of estimating approximately the turnover rate of histamine in various rat tissues. Their estimates agree in the main with ours and they are included in Table 17.

Table 17. *Turnover rate of histamine in various rat tissues*

Abdominal skin	Slow (A), (B), (C)
Ear	Slow (B)
Gastric mucosa	Rapid (A), (B), (C)
Kidney	Rapid (C)
Liver	Rapid (C)
Lung	Slow in one portion ⎱ (B) Rapid in a different portion ⎰
Small intestine	2 different rates, as in lung (B)
Tongue	Slow (B)

The estimates denoted (A) are those by Schayer; (B) ours; (C) Bjurö, Westling and Wetterquist.

In summary regarding turnovers: there are two extremes, slow turnover in mast cells (skin), rapid in the gastric mucosa. An important finding is that there are two modes of binding in predominantly non-mast-cell tissues, one portion having a rapid, and the other a slow, turnover. The inhibition experiments have thus disclosed three different kinds of binding. Each of the three distinct kinds of bound histamine is likely to have its specific physiological significance.

4. *Rebound, overshoot.* This phenomenon, illustrated in Figs. 16 and 17, which often occurs on discontinuing inhibition, occasionally persists for several days; it represents a state of elevated HFC maintained in the absence of drugs. It corresponds to a quasi-physiological state in which valid information can be sought on the physiological significance of 'nascent histamine', as will be discussed in the section of HFC and wound healing (Chapter 19).

Parenthetically, Fig. 17 illustrates the similarity in results obtained with the non-isotopic and the isotopic method. With the latter, whole-body HFC was determined before inhibition and at maximum inhibition of histamine formation (Kahlson, 1960).

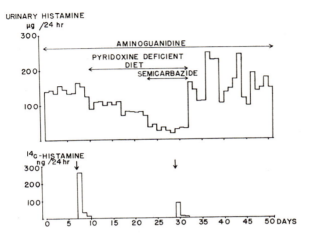

FIG. 17. Female rat fed on a histamine-free diet. *Upper section*, urinary excretion of free histamine; *lower section*, excretion of ^{14}C-histamine after subcutaneous injection of 220 µg ^{14}C-histidine. Semicarbazide was injected subcutaneously, 1st day 50 mg/kg twice daily, and 2nd day onward 75 mg/kg twice daily. Note rebound (overshoot) and similarity in results obtained with the non-isotopic and isotopic methods (Kahlson & Rosengren, 1968, *Physiol. Rev.*, **48**, 155).

5. *Origin of tissue histamine.* This subject has been fully discussed in Chapter 3. Our inhibition study provides evidence in support of the postulate that histamine from extracellular sources is not taken up and held by tissue cells under the conditions of this study. In the ear, following inhibition of histamine formation and a subsequent fall in its histamine content, at the fourth day of re-instituting adequate diet, resulting in complete restoration of HFC, the histamine content of the ear is only about half the normal. On strong and prolonged inhibition of histamine formation the histamine content falls even in the abdominal skin and small intestine (Table 14). At this stage there is a residual whole-body HFC, and histamine is available extracellularly, carrried by the

blood stream, as evidenced by its excretion in the urine. Nevertheless, this extracellular histamine is incapable of replacing intracellular histamine molecules lost continually in the turnover process.

α-methylhistidine

The feat of synthesizing α-methylhistidine on a μg scale was accomplished by Robinson (Robinson & Shephard, 1961a), a synthesis which, even with big factory capability available, has proved difficult. He prepared some potential inhibitors of decarboxylases and studied together with Shephard their ability to inhibit *in vitro* the decarboxylases of guinea-pig kidney and of transplantable rat hepatoma (Robinson & Shephard, 1961b; 1962). They used extracts. The two enzymes are now recognized, respectively, as DOPA-decarboxylase and histidine decarboxylase. In their work which appears mainly directed towards the elucidation of chemical structure–activity relationship, they found in the case of the kidney enzyme that the inhibitory action of α-methylhistidine was much weaker than that of α-methyl-DOPA, whereas with the hepatoma, α-methylhistidine produced substantial inhibition, although less so than did hydrazine compounds. These latter compounds were shown in our laboratory to be weak inhibitors *in vivo*, and as the original work by Robinson and Shephard was confined to determinations *in vitro* with extracts, we decided to study the inhibitory action of α-methylhistidine on histidine decarboxylase *in vivo* and *in vitro*, as well as the mode of action of this inhibitor. α-methylhistidine was provided by Merck, Rahway, New Jersey, through the kind office of Dr. Schayer.

Studies in vivo. Two types of experiment were carried out *in vivo*. In the first type, rats were fed on the standard diet, adequate in pyridoxine, and the urinary histamine was followed in the usual way. To see whether α-methylhistidine inhibits also the formation of 5-hydroxytryptamine (5-HT) (produced by the hydroxylation of tryptophan to 5-hydroxytryptophan and the subsequent decarboxylation of this amino acid) the urinary excretion of 5-hydroxyindolyl acetic acid (5-HIAA, the oxidation product of 5-HT), which is believed to reflect the rate of endogenous formation of this amine, was also determined. An experiment of this kind is illustrated in Fig. 18. α-methylhistidine $(4 \times 200$ mg/kg) was injected subcutaneously on four successive days. As a result, the daily urinary histamine was reduced by about 50%, i.e. the whole-body histamine formation was lowered to half of the normal. The smallest dose which gave a significant reduction was 4×25 mg/kg daily. It is seen from Fig. 18 that the inhibitory influence of α-methylhistidine disappears rapidly on discontinuing treatment. The excretion of 5-HIAA was not seemingly influenced even in daily doses as large as 4×200 mg/kg (Kahlson, Rosengren & Svensson, 1962). Rats under the influence of α-methylhistidine in the dose range employed did not exhibit discernible psychical or physical ill effects, in contrast to the rats receiving semicarbazide superimposed on the deficient diet.

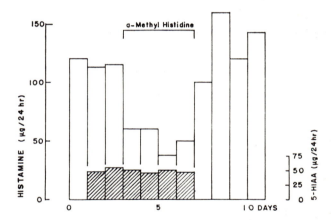

FIG. 18. The white columns represent amounts of daily excretion of free histamine in a female rat fed an adequate histamine-free diet. The hatched columns represent daily urinary excretion of α-HIAA. α-Methylhistidine was injected for a period of four days. Aminoguanidine was administered during the entire course of observations to prevent the inactivation of histamine by histaminase (Kahlson *et al.*, 1962, *Nature*, **194**, 876).

In another kind of experiment pyridoxine was omitted from the diet until a fairly steady lowered level of histamine formation was obtained. The purpose was to see whether α-methylhistidine would cause a further reduction in histamine formation, i.e. whether inhibition by this compound is due to interference with the histidine decarboxylase itself or merely with the co-enzyme pyridoxal phosphate. In the experiment represented in Fig. 19, the omission of pyridoxine for 24 days lowered the daily urinary histamine from 125 to 50 μg, a level which with small fluctuations was maintained for about a week. At this stage α-methylhistidine (4 × 200 mg/kg) on 4 successive days rapidly caused a further lowering to about 25 μg. The inhibitory effect of the drug persisted at this level during the 4 days of administration. On the first day after discontinuing α-methylhistidine the urinary histamine was restored to about 50 μg. At this stage an injection of 100 μg pyridoxal phosphate restored the level nearly to normal. The isotopic method, involving the injection of ¹⁴C-histidine and determining the urinary ¹⁴C-histamine in the normal state and under the influence of α-methylhistidine, gave similar results. The inadequate supply of α-methylhistidine at the time of our experiments prevented our studying its effect at higher concentrations on HFC.

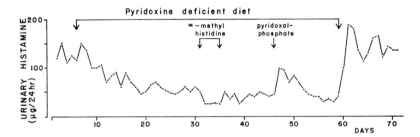

FIG. 19. Urinary histamine in a female rat on a pyridoxine-deficient diet and aminoguanidine. α-methylhistidine (4 × 200 mg/kg daily) injected subcutaneously for 4 days. At the arrow a single injection of 100 μg pyridoxial phosphate subcutaneously. Note rebound phenomenon on reinstituting an adequate diet (Kahlson *et al.*, 1963, *J. Physiol.*, **169**, 467).

Studies in vitro *and mode of action of* α-*methylhistidine.* In this part of the inhibitory experiments (1963) we used extracts of rat gastric mucosa. The influence of the concentration of substrate, co-enzyme and inhibitor was studied by altering the concentration of one of them at a time, other conditions remaining constant. α-methylhistidine inhibits histidine decarboxylase *in vitro*, as seen in Fig. 20. The degree of inhibition is highly dependent on the concentration of the substrate histidine and, as can be seen in

Fig. 20(c), inhibition is overcome by raising the concentration. Inhibition of α-methylhistidine is much less affected by the concentration of the co-enzyme, pyridoxal phosphate (Fig. 20(b)). The strong dependence on substrate concentration suggested to us that the mode of action of α-methylhistidine is competitive.

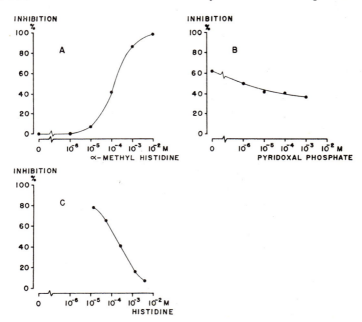

FIG. 20. Inhibition by α-methylhistidine of histidine decarboxylase from rat gastric mucosa *in vitro*. Effect of varying the concentration of α-methylhistidine in (a) pyridoxal phosphate in (b) and histidine in (c). When maintained constant the concentrations were: pyridoxal phosphate 10^{-5} M: histidine 5×10^{-5} M: and α-methylhistidine 10^{-4} M in (b), and $2 \cdot 5 \times 10^{-4}$ M in (c) (Kahlson *et al.*, 1963, *J. Physiol.*, **169**, 467).

Lovenberg, Weissbach and Udenfriend (1962), studying a high speed supernatant fraction from a guinea-pig kidney homogenate, discovered that it was capable of decarboxylating all the natural aromatic L-amino acids, and they designated this enzyme preparation as aromatic L-amino acid decarboxylase. On examining the characteristics of mammalian histidine decarboxylating enzymes, Weissbach, Lovenberg and Udenfriend (1961) showed that the activity of this guinea-pig enzyme extract was enhanced by benzene and strongly inhibited by α-methyl-DOPA. On present knowledge this enzyme should be recognized as a member of the DOPA-decarboxylase class, and the original name, introduced by Udenfriend and his colleagues, presumably was suggested as a useful introductory denomination which is likely to be replaced simply by DOPA-decarboxylase. In the study by Udenfriend and his colleagues (1961), in addition, they examined some characteristics of a supernatant fraction of mouse mast cells which are known to contain histidine decarboxylase. With the methods used, α-methylhistidine, even in the concentration 10^{-3} M, did not detectably inhibit histidine decarboxylase. It only later became known that this inhibitor is converted to α-methylhistamine by the action of histidine decarboxylase.

We surmised that the inhibitor became subjected to decarboxylation along with histidine, and hence we incubated extracts of rat foetal liver with α-methylhistidine. α-methylhistamine is biologically almost inactive and therefore not measurable by bioassay. A sensitive method developed by Conway (1950) for measuring small amounts of CO_2 was adapted to our needs by Siesjö (1962). As CO_2 is produced during amino acid decarboxylation, this method was used for the determination of the decarboxylation of both histidine and α-methylhistidine. We found that about as much CO_2 was evolved on incubation with α-methylhistidine as with histidine (Table 18). For details we refer to our publication of 1963.

It is now easy to see why the inhibitory action of α-methylhistidine, by necessity, was concealed from detection by Uden-

Table 18. Decarboxylation of α-methylhistidine and of histidine determined by measuring the CO_2 formation (μmole/g) during 3 hr incubation with rat foetal extracts from three different litters

Litter no.	Extracts alone	Extract + histidine	Extract + α-methyl-histidine
I	0·24	0·82	0·70
	0·29	0·77	0·64
II	0·57	1·02	1·43
III	1·96	2·40	2·20
	1·92	2·32	2·34

(Kahlson, Rosengren & Thunberg, 1963, *J. Physiol.*, **169**, 467.)

friend and his colleagues. They determined the residual histamine fluorometrically, whereby the α-methylhistamine formed is likely to simulate histamine, thereby concealing any inhibition that might have taken place. In the *in vitro* experiments by Robinson and Shephard (1962) histamine was assayed on the isolated guinea-pig ileum. Why they found relatively weak inhibition of histidine decarboxylase by α-methylhistidine remains unexplained. Indeed, the methodology for studying inhibition of histamine formation is intricate and is becoming ever more so; further examples of this will be referred to presently.

In sum regarding α-methylhistidine. Its strong and specific inhibition of histidine decarboxylase, *in vitro* and *in vivo*, was shown by both non-isotopic and isotopic methods in 1963. α-methylhistidine is decarboxylated to the inert α-methylhistamine and is therefore unlikely to effect very strongly inhibition of histamine formation *in vivo* unless this inhibitor is administered at brief intervals. α-methylhistidine is at present a safe means in identifying the presence of histidine decarboxylase activity and in distinguishing this activity from that of DOPA decarboxylase.

α-methyl-DOPA (β-3, 4-dihydroxyphenyl α-methylalanine)

In vitro this compound has been shown to inhibit strongly the kidney decarboxylase activity (in our terminology DOPA-decarboxylase) extracted from guinea-pigs (Mackay & Shephard, 1960; Werle, 1961; Weissbach, Lovenberg & Udenfriend, 1961), and from rabbits (Ganrot, Rosengren & Rosengren, 1961), but does not strongly inhibit decarboxylase extracts from mouse mast-cell tumours (Weissbach, Lovenberg & Udenfriend, 1961) or from rat foetal liver, i.e. histidine decarboxylase (Ganrot, Rosengren & Rosengren, 1961). In our *in vivo* experiments (1963) this compound did not significantly reduce histamine excretion in female rats, even when injected in the very toxic dose 4×400 mg/kg on 2 successive days. In a single experiment only did this dosage produce a definite reduction in histamine excretion, and, in addition a corresponding reduction in the excretion of 5-HIAA; this latter observation is in accord with the finding of Smith (1960) that α-methyl-DOPA is an inhibitor of 5-hydroxytryptophan decarboxylase *in vitro* and *in vivo*.

Hydrazine compounds

This group of carbonyl reagents, which engage pyridoxal phosphate, is known to inhibit amino acid decarboxylases *in vitro*. Early work in this field has been extensively reviewed by Clark (1963). Robinson and Shephard (1962) investigated the inhibitory effect *in vitro* of 15 compounds on histidine decarboxylases extracted from guinea-pig kidney and rat hepatoma. It is now agreed, as these workers seemingly did, that the former extract contains DOPA decarboxylase and the latter histidine decarboxylase. Their results are summarized in Table 19 (1962). Some of their observations conform with ours, and some do not. They found, as we did, that α-methylhistidine was a much weaker inhibitor of DOPA decarboxylase than was α-methyl-DOPA. Our results differ, in that in our *in vivo* studies the hydrazine derivatives tested were weak inhibitors and α-methylhistidine a strong inhibitor of histidine decarboxylase.

Table 19. Concentrations of various compounds required to produce 50% inhibition (C_{50}) of L-histidine decarboxylases

No.	Compound	$C_{50} \times 10^4$ M	
		GPHD	F-HepHD
1	Imidazole-4(5)-carboxylic acid	30	150
2	4(5)-Nitroimidazole-5(4)-carboxylic acid	65	300
3	Imidazole-4(5)-carboxyhydrazide	20	15
4	4(5)-Nitroimidazole-4(5)-carboxyhydrazide	6·5	0·75
5	L-Histidine hydrazide $1\frac{1}{2}H_2SO_4$	0·85	2
6	Hydrazine salt of 4	0·2	0·15
7	Hydrazine hydrate	0·35	0·1
8	DL-5-HTP	0·65	75
9	DL-α-Methyl-5-HTP	0·075	1
10	DL-α-Methylhistidine dihydrochloride	150	15
11	L-DOPA	0·2*	7·5
12	DL-DOPA	0·2	4·5
13	DL-α-Methyl-DOPA	0·01*	70
14	Catechol	1·8*	65
15	Salicylic acid	35	30

* Quoted from the results of Mackay and Shephard (1960).
(Robinson & Shephard, 1962, *J. Pharm. Pharmac.* **14**, 9.)

Further instances of entirely opposite results with histidine decarboxylase inhibitors will presently be discussed. Regarding this particular discrepancy there is a difference in procedure between the workers in Scotland and in Sweden. The former employed the substrate concentration 6.4×10^{-4} M whereby the inhibition by α-methylhistidine was much less than with histidine 8×10^{-5} M in our experiments (Fig. 20c). The Scottish group incubated at pH 6·8, we chose pH 7·4, a dissimilarity the consequence of which we cannot assess.

Isonicotinic hydrazine, credited with an inhibitory action on enzyme systems for which pyridoxal phosphate is a cofactor, is practically inactive on histamine formation *in vivo*. We gave thirteen daily doses of 100 mg/kg which resulted in a slight reduction in whole-body HFC, which subsided during the course of injections. It will be recalled from our experiments on dialysing tissue extracts that the cofactor of histidine decarboxylase appears to be firmly attached to the enzyme protein.

Ethyl hydrazine of podophylatic acid (SPI), administered intra-peritoneally in 0·9% NaCl in a dose of 10 mg/kg once per day for 3 days and as a single dose of 20 mg/kg was without effect on urinary free histamine excretion in four rats, investigated by Johnston and Kahlson (1967).

In the following we will describe and discuss experiments by these authors in which inconsistent results were obtained when inhibition was determined non-isotopically and isotopically. The α-hydrazino analogue of histidine (α-HH) will be discussed below under a separate heading.

Quercetin

It has been reported that this substance is an effective inhibitor of histidine decarboxylase both *in vitro* and *in vivo* (Smyth, Lambert & Martin, 1964). *In vitro* they employed extracts of gastric mucosa, liver and kidney of rats, and skin and mastocytoma of mice. Determinations were made with Schayer's ^{14}C-method, and for unexplained reason benzene was added to the incubation mixture, whereby histidine decarboxylase activity becomes lowered and that of DOPA decarboxylase enhanced, pH was 8·3. *In vivo*, ^{14}C-histidine was introduced into the ligated stomach, which was excised and assayed for ^{14}C-histamine. In addition urinary ^{14}C-histamine was measured after subcutaneous injection of ^{14}C-histidine, or ^{14}C-histamine was determined in the supernatant of excised tissues. Among the results recorded, a dose of 10 mg/kg rat administered intraperitoneally inhibited by 100% the overall histamine synthesis, as judged by measurement of ^{14}C-histamine in the urine after injection of ^{14}C-histidine.

We studied this seemingly promising compound and found that quercetin even in doses higher than those employed by the original investigators had no effect on the excretion of non-isotopic free histamine as assayed on the guinea-pig gut. Doses in the range 5–200 mg/kg were used in rats, single and repeated, and by subcutaneous and intraperitoneal routes. Administration of 1 g/day in the food for 2 days had also no effect on the daily output of free histamine in the urine. Next we employed the isotopic method and noted that amounts of ^{14}C-histamine excreted in the urine after injection of ^{14}C-histidine and quercetin at doses of 5 or 10 mg/kg were not significantly different from those obtained in the same rat after injection of ^{14}C-histidine alone. From these results Johnston and Kahlson (1967) concluded that quercetin was not an effective *in vivo* inhibitor of histidine decarboxylase. Attempts at explaining discrepancies between our observations and those reported by others will be given in concluding the intricate chapter on inhibition.

4-bromo-3-hydroxybenzyloxyamine (NSD-1055)

This compound was introduced by Smith and Nephew Research Ltd., Harlow, Essex, England, and the inhibition of histidine decarboxylase *in vitro* by this compound was investigated by Reid and Shephard (1963). They employed guinea-pig kidney as the source of what they refer to as non-specific histidine decarboxylase and the transplantable hepatoma F-HEP to provide the 'specific enzyme'. Their results are given in Table 20.

The table shows that the NSD-compounds differ from α-methyl-DOPA in that *in vitro* they are powerful inhibitors also of the 'specific enzyme'; the authors stated that these compounds are the most potent inhibitors of this enzyme described at that time. Reid and Shephard showed that *in vitro* the new compounds reacted rapidly with pyridoxal phosphate, and they expressed the view that the potency of the compounds as inhibitors is due in part to inactivation of the co-enzyme. Confirming our report in 1962, they re-established α-methylhistidine as an inhibitor of the 'specific enzyme'.

Table 20. Concentrations required of various compounds to produce 50% inhibition of L-histidine decarboxylases

Compound	$C_{50} \times 10^4$ M Non-specific enzyme	Specific enzyme
NSD 1024	0·0052	0·006
NSD 1034	0·0025	0·0098
NSD 1055	0·0027	0·0014
Imidazol-4(5)-ylmethoxyamine	0·065	0·0021
DL-α-methyl-DOPA	0·01	70
DL-α-methylhistidine	150	15

(Reid & Shephard, 1963, *Life Sci.*, Oxford, **2**, 5.)

Inhibition of histamine formation *in vivo* by NSD-1055 was studied in female Sprague–Dawley rats by Levine, Sato and Sjoerdsma (1965). The compound was injected intraperitoneally, 100 mg/kg, urine collection was begun immediately after the injection, the rats were scarified 24 hr later, and the histamine content in the heart, stomach and urine was determined. Their observations are given in Table 21. Noteworthy is the nearly 50% inhibition of histamine excretion as well as of histamine content of

Table 21. Histamine levels before and 24 hours after intraperitoneal administration of several compounds

Compound	Dose (mg/kg)	Heart* (μ/g)	Stomach* (μg/g)	Urine* (μg/24 hr)
None(control)		4·3 ±0·4 (35)	19·7 ±2·7 (20)	37·0 ±4·8 (31)
α-HH	100	3·2 ±0·4 (12, P < 0·001)	12·0 ±1·4 (8, P < 0·001)	22·4 ±6·7 (8, P < 0·01)
NSD-1055	100	2·9 ±0·3 (8, P < 0·001)	11·2 ±1·4 (8, P < 0·001)	19·8 ±2·2 (8, P < 0·001)
Methyldopa	100	4·1 ±0·3 (5, P > 0·05)	25·2 ±3·2 (5,P > 0·05)	41·1 ±7·1 (5, P > 0·05)
48/80	2†	2·5 ±0·4 (7, P < 0·001)	14·7 ±1·3 (7, P < 0·01)	75·7 ±13·4 (5, P < 0·01)

* Each value is expressed as the mean ±1 standard deviation; in parentheses are the numbers of animals studied and the probability of the chance occurrence of the differences from controls expressed as the P value.

† Administered in two doses (see text for details).

(Levine *et al.*, 1965, *Biochem. Pharmac.* **14**, 139.)

the stomach. After injection, the inhibition in the heart and stomach was maximal between 3 and 6 hr, the histamine levels began to return towards normal within 24 hr, and restoration was complete within 48 hr, as shown in Fig. 21. NSD-1055 was found to be toxic, in that 100 mg/kg injected every 12 hr killed each rat by the third day. Inhibition *in vitro* was studied with 'specific enzyme' extract (whole rat foetus) and with the 'non-specific' enzyme extract (kidney extracts *ad modum* Weissbach, Lovenberg and Udenfriend, 1961). From the results, which agree with those reported by Reid and Shephard (1963), it would appear that NSD-1055 is a strong inhibitor of both enzymes.

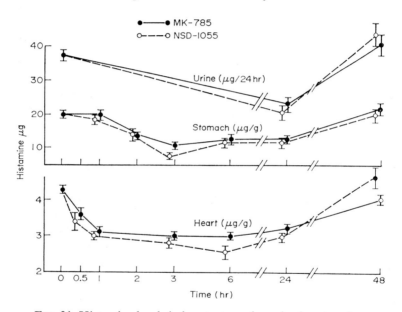

FIG. 21. Histamine levels in heart, stomach, and urine at various time intervals after i.p. administration of MK-785 (α-HH) and NSD-1055, 100 mg/kg. Each point represents the average of at least four observations; brackets designate standard errors of means (Levine *et al.*, 1965, *Biochem. Pharmac.*, **14**, 139).

Levine and his colleagues did inhibition studies with NSD-1055 in germ-free rats and noted, as we found in 1957, that the tissue histamine content was the same as in normal rats. Further, they recorded that in germ-free rats the inhibition of histamine formation was less in the stomach but, inexplicably, more pronounced in the urinary excretion than in normal rats in which histamine of intestinal origin has been alleged to contribute to histamine excretion. In their experiments injections of α-methyl-DOPA did not lower tissue or urinary histamine, indicating, as did our report on inhibition (1963), that in the rat DOPA-decarboxylase does not to any detectable degree contribute to endogenous histamine formation. They found that NSD-1055 injected subcutaneously in two doses of 100 mg/kg, 24 and 12 hr before the removal of tissues, did not inhibit histamine formation in peritoneal mast cells.

Levine (1966) gave NSD-1055 orally to normal humans, and reported that as a result the urinary excretion of histamine decreased, and that an otherwise occurring increase in urinary histamine following an oral load of histidine was prevented. Further, Levine (1968) administered NSD-1055 to two patients with Zollinger–Ellison syndrome and recorded a decreased basal acid secretion. This latter finding is the more striking since in our laboratory inhibition of gastric mucosal HFC did not diminish the acid secretory response to injected gastrin in rats (see Chapter 9). The effects of NSD-1055 reported in humans are perhaps due to factors other than inhibition of histamine formation, or, unlikely, histidine decarboxylase in humans is singularly susceptible to inhibition by NSD-1055.

Johnston and Kahlson (1967) followed the urinary histamine excretion over a period of about one month in rats and administered NSD-1055 at various intervals and in different doses. In one rat a dose of 100 mg/kg intraperitoneally (i.p.) was followed 6 days later by 100 mg/kg subcutaneously (s.c.) twice a day for 3 days, and after a further 12 days a single dose of 200 mg/kg i.p. was given. On no occasion was there any reduction in urinary histamine excretion as judged by bioassay. Two more rats were given a dose of 100 mg/kg (in one case i.p. and in the other s.c.) per day for 6 successive days; this treatment failed to have any effect on the urinary free histamine excretion.

In one rat, an i.p. dose of 200 mg/kg was given together with a s.c. injection of ^{14}C-histidine, and the radioactivity of the urine due to ^{14}C-histamine on the following 3 days was compared with that of a control period when no drug was given. Although no difference in non-isotopic histamine excretion was seen in the bioassay, there was a 54% reduction in the output of isotopic histamine when NSD-1055 was administered (Table 22).

Table 22. Effect of NSD-1055 and αHH on the urinary excretion of ¹⁴C-histamine after injection of ¹⁴C-histidine

Dosage (mg/kg)	Inhibitor test period			Control period		
	1st day	2nd day	3rd day	1st day	2nd day	3rd day
Rat M26						
NSD-1055	62	34	23	130	26	17
200 i.p.	(91)	(110)	(91)	(111)	(111)	(100)
Rat M25, α-HH	29	25	13	110	20	36
200 i.p.	(79)	(83)	(79)	(66)	(79)	(86)
Rat M29, α-HH	18	19	10	94	17	13
400 s.c.	(89)	(127)	(140)	(100)	(110)	(115)
Rat M30, α-HH	13	9	8	98	15	12
400 i.p.	(109)	(114)	(120)	(120)	(144)	(120)

Compounds were administered in the doses indicated in the table at the beginning of the 24 hr collection period on the first day of the inhibitor test period.
Figures without parentheses are ¹⁴C-histamine in counts/min per total daily urine volume; figures in parentheses are the corresponding bioassay values in µg free histamine/ 24 hr (i.p. = intraperitoneal; s.c. = subcutaneous)
(Johnston & Kahlson, 1967, *Br. J. Pharmac.* **30**, 274.)

We tested this compound *in vitro* by adding it in an amount equivalent to an *in vivo* dose of 200 mg/kg (assuming even distribution) to the incubation mixture, using abdominal skin, lung and gastric mucosa as test tissues. The results, shown in Table 23, indicate a very strong inhibition, exceeding 90%.

Table 23. Inhibition of histamine formation *in vitro* by αHH and NSD-1055

	HFC, ng histamine formed/g tissue		
	No addition	+α-HH 0.5×10^{-3} M	+NSD-1055 0.5×10^{-3} M
Abdominal skin	50	2	1
Lung	1220	144	7
Gastric mucosa	910	14	11

(Johnston & Kahlson, 1967, *Br. J. Pharmac.* **30**, 274.)

Reilly and Schayer (1968) obtained nearly 100% inhibition *in vitro* of rat stomach histidine decarboxylase (Table 24) with much lower concentrations of NSD-1055 than those employed in our experiments (Table 23). Besides, they introduced a new inhibitor of histidine decarboxylase, *p*-toluensulphenyl hydrazine (PTSH), which we tried and found toxic in mice. Further, they found NSD-1055 rather ineffective *in vivo*, as judged by the urinary excretion of ^{14}C-histamine in rats injected with ^{14}C-histidine. They proposed that *in vivo* NDS-1055 and α-HH do not reach the histidine decarboxylase containing sites. Johnson (1969) found that NSD-1055 was ineffective in reducing endogenous histamine levels and did not affect the kinetics of decline of intravenously injected ^{3}H-histamine which he assumes to equilibrate with non-mast-cell histamine in the rat.

Table 24. Inhibition of rat stomach histidine decarboxylase *in vitro* by PTSH, α-HH and NSD-1055

	% inhibition at various concentrations		
Inhibitor	2×10^{-6} M	5×10^{-7} M	1.25×10^{-7} M
PTSH	97	86	45
α-HH	98	87	42
NSD-1055	100	89	69

(Reilly & Schayer, 1968, *Br. J. Pharmac.* **34**, 551.)

Conclusions regarding NSD-1055. This compound is of dubious avail as inhibitor in studies of histamine formation *in vivo*. It is not useful in discriminating between histidine decarboxylase and enzymes of the DOPA-decarboxylase class because both enzymes are inhibited to about the same degree. Notwithstanding, inhibitory studies with this compound have brought forth significant knowledge and posed new problems:

(1) Whole-body histamine formation (excretion of non-isotopic histamine) may remain unaltered even in instances where the inhibitor, as in our experiments but less consistent in those by Reilly and Schayer, inhibits the decarboxylation of injected ^{14}C-histidine; i.e. the dilemma arises as to whether the non-isotopic or the isotopic method should be resorted to in investigating inhibition of histamine formation *in vivo*.

(2) Seemingly equal concentrations of an inhibitor may be strongly active *in vitro* but less so or totally ineffective on histamine formation *in vivo*. The discrepancy in results obtained *in vitro* and *in vivo* could be explained on the assumption of a more rapid elimination of the inhibitor *in vivo*.

α-Hydrazine analogue of histidine (α-HH)

This compound was synthesized by Merck, Sharp & Dohme, Rahway, U.S.A., who also produced a batch of α-methylhistidine to be donated to various workers in the field. Inhibitory properties of α-HH were first investigated by Levine, Sato and Sjoerdsma (1965). Among four compounds studied on tissue extracts *in vitro* (Table 25) only α-HH was reported to be a more potent inhibitor of the 'specific' (whole foetal rats) than the 'non-specific' (guinea-pig kidney) enzyme.

Table 25. Inhibition of histidine decarboxylase *in vitro* by four compounds

Compound	Concentration ($M \times 10^5$) producing 75% inhibition*	
	Specific enzyme	Non-specific enzyme
α-HH	7	80
NSD-1055	0·04	0·009
Methyldopa	>100	0·08
α-Methyldopa-hydrazine	20	0·09

* Numbers represent concentrations of each compound that produced 75% inhibition of activity, estimated as described in the text and multiplied by 10^5. Thus, low numbers indicate high potency as inhibitors. Each number represents the average of values obtained in at least two experiments.

(Levine *et al.*, 1965, *Biochem. Pharmac.* **14**, 139.)

On injecting 100 mg/kg i.p. and determining after 24 hr 'specific histidine decarboxylase' activity in the heart and stomach, as well as histamine excretion, the degree of inhibition recorded was about the same as in these authors' corresponding experiments with NSD-1055. When α-HH, 100 mg/kg, was injected every 12 hr for 4 days, histamine excretion decreased as with NSD-1055 (Fig. 22); with α-HH no toxicity was apparent during 4 days of study.

We investigated α-HH *in vitro* and *in vivo* as described for NSD-1055. *In vitro* we used three different rat tissues finely cut. The results, shown in Table 23, indicate an inhibition of about 90–100% *in vitro*; Reilly and Schayer (1968) noted strong inhibition of the stomach enzyme with α-HH in a much lower concentration (Table 24) which we at that time (1967) unfortunately did not try.

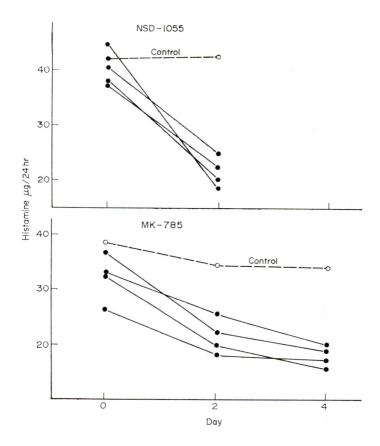

Fig. 22. Urinary excretion of histamine during administration of NSD-1055 and MK-785 (α-HH), 100 mg/kg i.p. every 12 hr. Each line joins points representing values obtained in a single rat (Levine *et al.*, 1965, *Biochem. Pharmac.*, **14,** 139).

In vivo we investigated inhibition by α-HH in various doses: a single i.p. injection of 100 mg/kg and 200 mg/kg daily for 3 days had no effect on urinary histamine output as judged by bioassay. Also with 100 mg/kg injected i.p. every day for 6 days no reduction was seen. Next, the excretion of ^{14}C-histamine after injection of ^{14}C-histidine was measured with and without administration of α-HH in doses of 400 mg/kg (i.p.), 400 mg/kg (s.c.) and 200 mg/kg (i.p.) being used. The results, shown in Table 22, indicate inhibitions of 87%, 81% and 75% respectively, although the concurrent bioassay of non-isotopic urinary histamine showed no inhibition.

Having seen that NSD-1055 and α-HH inhibited the decarboxylation of injected ^{14}C-histidine more effectively in experiments in which these compounds were injected concurrently with, or shortly after, the injection of ^{14}C-histidine, we decided to investigate whether repeated injections of these inhibitors at shorter intervals over 2 days would show inhibition of histamine formation with the non-isotopic method. Rats were injected with α-HH, twice with 100 mg/kg (s.c.) on the first day and a further four times with 50 mg/kg on the second day. Only on the second day, a slight but significant reduction in urinary histamine excretion occurred. Similarly, four rats were given NSD-1055 for 2 days, three times 100 mg/kg (s.c.) on the first day, three times 200 mg/kg (i.p.) on the second day. As with α-HH, the reduction in urinary histamine was insignificant during the first day, but on the second day the four rats displayed respectively 35%, 44%, 50% and 63% inhibition as expressed in the urinary histamine. It thus appears that even under the influence of large doses of these compounds injected at brief intervals the inhibition of whole-body HFC is inconsistent and moderate. Consequently, in these instances results with injected histidine is a dubious measure of the degree of inhibition of histamine formation.

Similar isotopic *in vivo* experiments have been carried out by Reilly and Schayer (1968). They injected ^{14}C-histidine (s.c.) in female rats, pretreated with inhibitors, collected urine overnight and determined the excretion of ^{14}C-histamine. Their results, mainly similar to ours, are shown in Table 26. With α-HH they obtained about 50% inhibition; NSD-1055 was even less active.

In mice given PTSH or α-HH plus ^{14}C-histidine, Reilly and Schayer found the amount of ^{14}C-histamine in the skin four days later drastically reduced, whereas NSD-1055 had a much weaker inhibitory effect on ^{14}C-histamine formation in the skin. Since mouse skin contains large numbers of mast cells, Reilly and Schayer believe that the effective inhibitor penetrates these cells *in vivo* and inhibits histidine decarboxylase therein. From their results in mice they infer that *in vivo* PTSH and α-HH and to a much lesser extent, NSD-1055, inhibits stomach histidine

Table 26. Effect of histidine decarboxylase inhibitors on (^{14}C)-histamine in urine of female rats injected with (^{14}C)-L-histidine

Treatment	(^{14}C)-Histamine, d.p.m. in total urine		
Saline (control)	2560	2960	3150
PTSH	1000	1270	1610
α-HH	1370	1440	1740
NSD-1055	2210	2480	2650

First injection, 0 time, saline or inhibitor, 25 μM/100 g body weight combined with aminoguanidine, 1 mg/100 g body weight; second injection, 40 min, (^{14}C)-L-histidine 3·0 μc s.c.; third injection, 2 hr, equivalent to one-half of dose used in first injection. Single estimations on pooled urines from two rats.

(Reilly & Schayer, 1968, *Br. J. Pharmac.*, **34**, 551.)

decarboxylase. *In vitro*, by contrast, NSD-1055, at the lowest concentrations tested by them, was a more effective inhibitor of rat stomach histidine decarboxylase than either PTSH or α-HH.

Isotopically, *in vivo* in rats, Reilly and Schayer obtained only insignificant inhibition with α-HH, whereas isotopically we found 75–87% inhibition (Johnston & Kahlson, 1967). We injected larger doses of this inhibitor than Reilly and Schayer who believe that large doses of any hydrazine derivative, even without contact with the histidine decarboxylase holoenzyme, can inhibit histamine formation by trapping the loosely held co-enzyme pyridoxal phosphate. The firmness of the binding of this cofactor *in vivo* would be difficult to assess. In our study in 1963 of rat gastric mucosa extract it was difficult to remove pyridoxal phosphate by dialysis. Further, injections of desoxypyridoxine and isonicotinic hydrazine, both believed to be pyridoxine 'trappers', did not on injection inhibit whole-body histamine formation in rats. On the other hand, pyridoxine-deficient diet, instituted over weeks, inhibited whole-body histamine formation by about 90%. Indeed, *in vivo*, a few days of pyridoxine-deficient diet substantially reduced the whole-body HFC. From this we conclude that *in vivo* the fully active molecules of histidine decarboxylase rapidly disappear, i.e. the turnover rate of this enzyme is rapid. Further evidence of a high rate of synthesis of histidine decarboxylase is given by the phenomenon of 'overshoot', and the steep increase in gastric mucosal histidine decarboxylase activity on injecting gastrin. As to the

D

mode of action of hydrazine derivatives *in vivo*, this action, indeed, appears not clearly explained.

Chloramphenicol

Experiments employing this compound carried weight at an earlier stage of attempts to elucidate the origin of tissue histamine. Wilson (1954) showed in clear-cut experiments in female rats that chloramphenicol administered for 4 days in succession caused a decrease in the excretion of free histamine to 45% of normal on the third day of administration, whereupon the excretion gradually returned towards the control value which it even exceeded before finally returning to normal. (Perhaps the phenomenon which we refer to as 'overshoot'.) Aureomycin, phthalyl sulphathiazole and penicillin produced effects on the histamine excretion similar to those by chloramphenicol. Wilson compared the changes in histamine excretion with changes known to be produced by the antibiotics in the bacterial flora of the intestine. From these observations and arguments he concluded that under physiological conditions, in the normal rat, about half of the histamine excreted in the free and conjugated forms is produced as a result of bacterial activity within the lumen of the intestine, whence it is absorbed, and finally excreted in the urine.

In our study of germ-free rats, discussed in Chapter 3, we made observations which were not consistent with the view that a considerable portion of urinary histamine is of intestinal-bacterial origin. We thought that the pertinent circumstances in rats may differ in the various laboratories concerned. We therefore decided to test chloramphenicol. In the 1967 study the daily output of urinary non-isotopic histamine was followed by bioassay in eight female rats treated with chloramphenicol as indicated in Table 27. This table gives estimated inhibitory effects, comparing the minimum post-treatment value with the pre-treatment level of urinary histamine in each individual rat. The inhibitory effects are small.

Assays of ^{14}C-histamine excretion after injection of ^{14}C-histidine were made in six of the rats detailed in Table 27, where inhibitory effects, as judged by comparison with a pre-treatment control period, are shown. Equally, by this method, the effects of chloramphenicol were small. We made *in vitro* determinations in various tissues of twenty Sprague–Dawley rats which had been given

Table 27. Inhibitory effect of chloramphenicol on urinary histamine as measured by bioassay and by excretion of ^{14}C-Histamine after injection of ^{14}C-Histidine

Rat dosage (mg)	Treatment No. of days	Inhibitory effect % Biossay	Isotopic method
1 32 in food/day	1	15	10
2 32 in food/day	3	0	20
3 32 in food/day	3	30	10
4 32 in food/day	4	20	
5 16 s.c. twice/day	1	30	0
6 16 s.c. twice/day	3	50	20
7 16 s.c. twice/day	3	30	30
8 16 s.c. twice/day	4	37	

The figures for inhibitory effects as measured by the bioassay and isotopic method cannot be directly compared in each rat, since both were not necessarily calculated for the same days. The inhibitory effect is expressed as per cent of non-isotopic histamine (bioassay) excreted and of ^{14}C-histamine excreted as compared with the pre-treatment values. (Johnston & Kahlson, 1967, *Br. J. Pharmac.*, **30**, 274.)

chloramphenicol for 4 days, one 32 mg in the food and the other 16 mg twice per day s.c. The HFC of the various tissues did not significantly differ from non-treated controls.

Various antimitotic agents

Since a connexion between histidine decarboxylase activity (nascent histamine) and certain types of rapid tissue growth and mitotic rate has been established, we tested some antimitotic agents, developed for cancer therapy, for the inhibition of histamine formation, in female rats in our 1967 study.

1. *1,2-dimethyl-4-(p-carboxy-phenylazo)-5-hydroxybenzene* (*CPA*) given in the food at a dose of 90 mg/day for 10 days produced a fall of about 50% in urinary histamine excretion immediately, but the output began to rise again after 1–2 days, despite continued administration of the drug.

2. *1,2-dichloro-4-benzene-sulphonamido-5-nitrobenzene* (*DCBN*), given in the food at a dose of 7·5 mg/day for 10 days produced no fall in urinary output of free histamine.

3. p-*di-(2-chloroethylamino)-L-phenylalanine* (*Melphalan*), given at a dose of 5 mg/kg, i.p. in olive oil twice per day for one day produced a reduction of about 50% in urinary free histamine about

5 days after administration. Toxic effects occurred which rendered the compound unsuitable for use as an *in vivo* inhibitor of HFC.

4. *NN-di-2-chloroethylaminobenzene* (*aniline mustard*), when given at a dose of 50 mg/kg, i.p. in olive oil twice per day for one day, caused no significant change in histamine excretion.

5. *Cyclophosphamide* (*Endoxan*), was administered i.p. at a dose of 50 mg/kg two times per day for one day which caused maximum reduction of 40–50% in histamine excretion. No reduction was seen with 20 mg/kg once per day for 9 days. *In vitro* experiments failed to demonstrate any effect of cyclophosphamide on the rate of histamine formation at a concentration of 10^{-4} M, which corresponds to a dose of 100 mg/kg *in vivo*, assuming even distribution in the animal. It appears possible that a reaction occurring *in vivo*, but not under the *in vitro* conditions used, converts this compound to a relatively weak inhibitor of histidine decarboxylase.

CONCLUDING REMARKS ON INHIBITION

It appears disturbing that so much of the work published on inhibition of histamine formation, i.e. inhibition of histidine decarboxylase, has given incompatible results, and that the reasons for the inconsistencies obtained by various investigators cannot always be clearly seen. These discrepancies are likely to become narrower by agreement as to the nature of the enzyme to be measured and by what methods. As to methods, alas, the experiments *in vivo* have disclosed that the two methods employed in our laboratory, the isotopic and the non-isotopic methods, gave different and seemingly incompatible results with the inhibitors NSD-1055 and α-HH. These latter results are the more disquieting, since, as described in Chapter 3, these two *in vivo* methods generally give concordant results. At the present stage the following propositions and suggestions appear reasonably well founded:

1. The female rat, fed and treated as described in Chapter 3, is particularly well suited for studies of inhibition of histamine formation, as in this animal a large proportion of the histamine formed in the body is excreted as free histamine which can easily be assayed. Further, histidine decarboxylase, to the best of present knowledge, is the enzyme largely, or solely, instrumental in the formation of histamine *in vivo*. This view is based on the observation that in the rat the specific inhibitor of histidine decarboxylase,

α-methylhistidine, decreases the formation and excretion of histamine. By contrast, α-methyl-DOPA, which inhibits histamine forming decarboxylases other than histidine decarboxylase, does not appreciably inhibit histamine formation, as judged by the urinary excretion of the amine (Werle, 1961, our study, 1963).

2. In attempts to understand the difference between the non-isotopic and the isotopic methods and to assess the significance of results obtained with them, the following considerations appear helpful. In the first place, it should be emphasized that the isotopic method determines the decarboxylation, in this case the inhibited decarboxylation of an additional injected amount of the substrate, admittedly a very small amount indeed. The non-isotopic method determines histamine formed from histidine in its normal set-ting—in terms of physiology. Besides, we have to consider the prospective turnover rates of ^{14}C-histamine formed in various tissues from injected ^{14}C-histidine. The lifetime of histamine in various tissues is stated in Table 17. In the rat a large part of the histamine formed and excreted is of gastric-mucosal origin. Follow-ing a single injection of ^{14}C-histidine most of the ^{14}C-histamine formed in the pools with rapid turnover is excreted within 24 hr. On establishing inhibition a steep fall in excretion of ^{14}C-histamine from the 'rapid pools' occurs whereas long lifetime ^{14}C-histamine formed in 'slow pools' is excreted over a long period of time (Table 17). Pictorially, with the isotopic method there is a peak of ^{14}C-histamine formation and excretion and a subsequent slowly declining plateau of excretion. In the non-isotopic method, on the other side, there is no equivalent to this spike which could be cut down by enzyme inhibition, because in the normal state the intracellular histidine concentration is steady. Nevertheless, it is difficult to see why certain compounds which display substantial inhibition with the isotopic method fail to do so non-isotopically. The discrepancy between the two methods could be resolved by the easy proposition that the inhibitory compound more readily makes effective contact with the injected than with the substrate in its natural disposition. Further work in the complex field of inhibition of histamine formation may well disclose that in resolving the seemingly paradoxical results obtained with the isotopic and the non-isotopic methods, also other circumstances than the above discussed will have to be considered. What really matters is awareness of the fact that in inhibition studies the isotopic and the

non-isotopic methods have given divergent results with certain inhibitors, and that in such studies the two methods are in a way inherently different.

3. Determination of HFC in excised tissues of rats submitted to inhibition treatment may lead to over-rating the degree of inhibition prevailing *in vivo*.

4. Inhibition of histidine decarboxylase is one of the most helpful means in studies of some aspects of the physiology of histamine. In order to serve this purpose the means of inhibition must be strongly effective *in vivo*.

5. Finally, 100% inhibition of histamine formation *in vivo* is not likely ever to be achieved because of the feed-back relation between enzyme and end product.

9

THE FUNCTIONS AND STATUS OF HISTAMINE IN THE GASTRIC MUCOSA

THE presence in the gastric mucosa of large amounts of histamine, its powerful stimulatory action on the hydrochloric acid secreting cells when injected, and the demonstration of a lowered mucosal histamine content associated with excitation of acid secretion, this triad of firmly established facts, has, depending on the scrutineer's inclination, been taken either as a strong, or an inadequate, evidence of a role for histamine in exciting acid secretion. On reasoning along these lines it is overlooked that even in the fasting state gastric mucosal histamine is continuously formed, and, owing to its brief lifetime, histamine steadily leaves the mucosa, is 'mobilized' to enter the blood stream, and is in part excreted as such or as its metabolites. Yet, in some species there is no inter-digestive secretion. That is to say, histamine can pass through the mucosa and on its course to the blood stream evade stimulatory functional contact with the secreting cell.

On feeding, a steep acceleration of mucosal histamine formation ensues and the mucosal HFC persists elevated for hours, as does the acid secretion. It will be explained on the following pages that this phenomenon is strictly associated with the process of exciting acid secretion. The HFC elevation on feeding is preceded by a slight fall of preformed mucosal histamine (Fig. 25). This latter phenomenon deserves bringing into prominence: feeding is the only hitherto recognized physiological circumstance in which preformed tissue histamine has been shown to be mobilized. By pharmacological procedures, e.g. injection of gastrin preparations, a brisker fall can be obtained.

As to its function, there are presumably different kinds of mucosal histamine. The gastric mucosa is a site of very rapid cell renewal (LeBlond & Walker, 1956) and, implicitly, high rate of protein synthesis. In this kind of restorative processes histamine formation often proceeds at high rates (Chapters 19 and 22). A fraction of gastric mucosal histamine, derived from what we refer to as 'basal mucosal HFC', unrelated to secretion, may be associated with the anabolic processes of cell renewal and protein synthesis.

The problem of gastric histamine is, indeed, becoming increasingly complex. Half a century has passed since the discovery that injected histamine stimulates acid secretion and, though thousands of published records since have not resolved the pertinent problem of its function in the mucosa, they have, rewardingly, generated fresh problems pertaining to various aspects of the physiology of histamine. The history of experiments and views on gastric histamine will now be described as they progressed over the last three decades.

Search for histamine in blood from the stomach on exciting secretion

Transmission by the gastric vagus is cholinergic. Dale and Feldberg (1934) stimulated the vagi in dogs and found easily detectable amounts of acetylcholine in the venous blood from the stomach when eserine had been injected. Acetylcholine appeared in the venous effluent even in the absence of electrical vagus stimulation. The nature of this experiment did not enable the site of liberation of acetylcholine to be established, i.e. whether from the mucous membrane or the muscle coats. Pertinent to later discussions is their finding that a vagal cholinergic tone exists in the stomach of the dog, even under anaesthesia. This work was extended by Paton and Vane (1963), in a study of the responses of isolated stomachs of guinea-pigs, kittens, rats and mice, to electrical stimulation of the vagus nerves. Their results are in favour of acetylcholine being the sole transmitter at both the pre- and postganglionic endings in the stomach. According to their interpretation, histamine would excite cholinergic nerve terminals not derived from the vagus and presynaptic to the final effector ganglion, i.e. nerve terminals of an intercalated ganglion cell. From their experiments they found no support for the concept that histamine is released directly by nervous tissue.

Babkin (1938), in a lecture, perhaps with the work by Dale and Feldberg in mind, speculated freely, that stimulation of the vagal supply to the gastric glands may liberate acetylcholine which, in turn, would release histamine from some site in the vicinity of the oxyntic cell. Babkin's *magnum opus, Secretory Mechanism of the Digestive Glands* (1950) contains these pertinent prophetic observations: 'Histamine is deposited in the gastric mucosa in an inactive state. A special mechanism is necessary to enable it to act on the parietal cells.'

MacIntosh (1938), working in Babkin's laboratory, tested the speculation regarding vagal histamine release. He stimulated the vagi in anaesthetized dogs and determined the histamine content of gastric venous whole blood. On vagal stimulation no increase was observed. MacIntosh remarked that in his experiments an increase in histamine content possibly might have been detected if plasma had been examined.

Emmelin, Kahlson and Wicksell (1940) carried this type of experiment further by examining plasma. In dogs under chloralose, and in decerebrated cats, both vagal trunks above the diaphragm were stimulated for 30 minutes and arterial and venous gastric blood collected before, during and after vagal stimulation. In each of the four dogs and in four of the seven cats examined, a significant increase in the histamine concentration of gastric venous plasma was observed. The record of these experiments which was submitted for publication in the then *Skandinavisches Archiv für Physiologie* contains this paragraph: 'It is remarkable that the gastric secretory nerves must be stimulated about thirty minutes until the histamine concentration in the gastric venous plasma rises to a demonstrable amount. It is further remarkable that the output of histamine from the stomach wall continues during a long period after the end of stimulation.' While the report was in press, the authors repeated the experiments with embarrassingly inconsistent results. They also became disturbed by the observation in the report that the increase in venous plasma histamine persisted for up to 45 minutes after the end of stimulation, and they felt therefore obliged to withdraw the report from publication. It will later be shown that vagus stimulation greatly increases the rate of histamine formation in the gastric mucosa and that the elevated HFC persists for several hours.

Babkin's original speculation, once more, attracted experimental exploration by Irvine, Ritchie and Adam (1961). They stimulated the vagi centrally in dogs by injecting insulin. Although as a result the stomach secreted, there was no detectable increase in histamine content of venous gastric plasma by the method employed. In anaesthetized cats Blair, Dutt, Harper and Lake (1961) determined the concentration of free histamine in arterial plasma during gastrin infusion and during control periods. The arterial plasma histamine concentration during gastrin stimulation was not significantly different from that obtained during control conditions. Neither did these investigators (1962) find evidence of increased plasma histamine content of the gastric venous blood in the presence of gastrin stimulation. They concluded that it was possible that the method employed might not have been of adequate sensitivity to detect small amounts of histamine released.

In experiments of this type emphasis should be laid on the inadequacy of the method available at that time in determining blood or plasma histamine concentrations. There is no place for doubt that part of the histamine formed on exciting acid secretion, by vagal or gastrin stimulation, will enter the gastric venous blood. Applying a fairly sensitive method devised by Adam, Hardwick and Spencer (1957) for estimating histamine, Caridis, Porter and Smith (1968) observed in fasting human subjects an increased histamine content of the venous blood draining the body of the stomach while acid was secreted in response to intravenously administered pentagastrin.

Histamine in the gastric juice

After the failure to obtain consistent and convincing results with determinations of venous plasma histamine, Emmelin and Kahlson (1944) embarked on an extensive study of the histamine content in the gastric juice. They excited gastric secretion in anaesthetized cats and in dogs provided with a Pavlov pouch. The means of stimulation were: sham-feeding and feeding, electrical stimulation of the vagus, intravenous injections of histamine, priscol, eserine, gastrin, and acetylcholine injected into the cerebral ventricles. Whatever the mode of excitation, the histamine content of the gastric juice was about the same in each cat and in the range 25–50 μg/l. in the total of six cats. In four sham-fed dogs the histamine content of the juice from the pouch was in the range 10–20 μg/l. in the cephalic phase and 30–50 μg/l. when meat was introduced in the main stomach (gastric phase). In seven anaesthetized dogs, electrical stimulation of the vagus resulted in secretion of gastric juice with a histamine content of 8–120 µg/l. At that time it was believed that the histamine present in the gastric juice originated from histamine released in the mucosa when it was stimulated to secrete. The gastrin extract injected in cats was prepared as described by Komarov (1942). This study by Emmelin and Kahlson has perhaps some historical interest as the first recorded attempt to incite histamine release from the gastric mucosa by gastrin injection. They concluded (1944): 'Histamine free preparations of "gastrin" cause a liberation of histamine from the fundic mucosa as judged by the occurrence of histamine in the acid juice.'

Several subsequent workers have determined the histamine content of gastric juice under various circumstances. Their findings have been reviewed by Code (1956), and by Ivy and Bachrach (1966). To the discourse of these reviewers we propose to add the reflection that histamine in the gastric juice occurring on stimulation may, wholly, or at least in part, be contained in dissociated fragments from decaying mucosal cells.

A study by Born and Sewing (1967) in which radioactive histamine was infused should be mentioned. Their object was to establish whether there was a correlation between the histamine content of the plasma and that of the gastric juice and how the histamine concentration in the juice was affected by the intravenous infusions of histamine. In anaesthetized cats with a gastric fistula ^{14}C-histamine was infused intravenously for 3 hours. During the course of infusion there was no correlation between the rate of gastric secretion and the concentration of histamine in either plasma or gastric juice. Born and Sewing suggest that the presence of histamine in the gastric juice is not essential for gastric secretion, and that the presence of histamine in the juice is incidental and due to simple diffusion.

Content and distribution of histamine in the stomach wall

With the object of testing Edkins' gastrin theory, Gavin, McHenry and Wilson (1933) decided that 'in view of this similarity between "gastrin" and histamine it seemed advisable to study the distribution of the amine in various parts of normal stomachs'. Extracts were prepared by the hydrochloric acid procedure employed by Best and McHenry (1930). In dogs they found about 80% of the amine in the fundic mucosa but only a little in the pyloric mucosa. From this distribution of secretagogue material they concluded that the result did not support the findings of Edkins.

In connexion with the investigation of histamine in the gastric juice, Emmelin and Kahlson (1944) also determined the histamine content of the gastric mucosa in dogs and cats. The fundus region contained about twice as much as the pyloric region and throughout there was more histamine in the canine mucosa.

The first attempt to establish the differential distribution of histamine in the mucosa was made by Douglas, Feldberg, Paton and Schachter (1951). In the dog they found about twice as much histamine in the fundus as in the pyloric region. Moreover, the muscularis mucosae contained more histamine than the glandularis mucosae, and the submucosa which contains neither glandular nor muscular tissue was relatively rich in histamine. Finally, in the small intestine the histamine content was high; in the duodenum it was even higher than in the fundus of the stomach. This attempt was carried further by Feldberg and Harris (1953) who established 'histamine profiles', with one peak in the region of maximum concentration of parietal cells (as defined by Hollander, 1952) and a second peak in the region of the muscularis mucosae. In man, mucosal histamine profiles have been constructed by Smith (1961). He found, as did Feldberg and Harris, two peak zones of histamine for the body of the stomach, the highest in the region of the parietal cells.

HISTAMINE FORMATION IN THE GASTRIC MUCOSA

The gastric mucosa of all the species studied so far—man, dog, cat, rat, mouse and frog—form histamine from histidine by the agency of histidine decarboxylase (Kahlson, Rosengren, Svahn & Thunberg, 1964). Gastric mucosal histamine formation has been most extensively studied in the rat for these reasons: histidine decarboxylase activity is high, the rat stomach lacks means to inactivate histamine formed therein, in the female rat a large part of the histamine formed in the stomach is excreted as free histamine on which account changes in gastric mucosal HFC are reflected in corresponding changes in the urinary excretion of histamine. The rat is becoming increasingly useful, or perhaps indispensable, in this particular field of study because the unanaesthetized rat has recently been found most suitable in quantitative studies of the gastric secretion (Svensson, 1970*a*). In the rat a large proportion of the whole-body histamine formation takes place in the stomach. On removal of the stomach the urinary excretion of histamine falls to less than half the pre-operative value in the female (High, Shephard & Woodcock, 1965). This should not be taken to show that this great fall is due solely to the absence of histamine formation in the stomach. It will be explained later, that the contribution of the bone marrow in whole-body histamine formation is so large that half of the whole formation cannot reside in the stomach.

It has been stated by some commentators that mucosal histamine formation pertains only to the rat, but that in other species mucosal histidine decarboxylase is likely to lack physiological significance because the activity is so low compared to that in the rat. This statement appears to be based on the assumption that the parietal cells are equally sensitive to histamine in different species. In fact, the sensitivity to secretory stimulation by histamine is widely different in the species investigated, as is shown in Table 28. Lorenz, Halbach, Gerant and Werle (1969) found mucosal histidine decarboxylase in each species investigated, as we did. The values as determined by them are summarized in Table 29. The

Table 28. Mucosal HFC and histamine content and amount of histamine which on injection evokes definite acid secretion

	HFC (µg/g) (fasting)	Histamine content (µg/g)	Histamine dose (mg/kg as base) eliciting definite acid secretion
Rat	2·0	81	1·0
Mouse	0·86	14	0·5
Man	0·04	44*	0·004†
Cat, adult	0·04	28‡	0·02
young	0·16	30	—
Dog	0·02	85§	0·003†
Guinea-pig	0·06	7	0·01
Frog	0·08	7	—

* Smith (1959).   ‡ Haeger & Kahlson (1952*a*).
† Feldberg & Schilf (1930).   § Emmelin & Kahlson (1944).
(Kahlson, Rosengren, Svahn & Thunberg, 1964, *J. Physiol.* **174**, 400.)

high activity found in the antrum of some species is noteworthy. They underline that the enzyme activity of the pig mucosa equals that found in the hamster placenta by Rosengren (1966) which is the highest so far recorded. The method employed by the Munich workers was non-isotopic and details should be studied in their report. The values given in Table 29 are at variance with ours in Table 28.

Table 29. Activity of the specific histidine decarboxylases in crude extracts of the gastric mucosa

Species	n	Fundus	n	Corpus	n	Antrum
		Histidine decarboxylation in pmoles histamine formation/min and mg protein				
Man	7	$10\cdot6\pm2\cdot6$	12	$7\cdot1\pm1\cdot4$	3	$4\cdot9\pm1\cdot2$
Monkey	3	$18\cdot7\pm7\cdot7$	5	$4\cdot6\pm1\cdot8$	3	$7\cdot2\pm1\cdot7$
Dog	2	$38\cdot4$	2	$46\cdot5$	2	26
Cat	1	$7\cdot4$	—	—	—	—
Pig	4	$202\quad\pm105$	4	$60\quad\pm14$	4	$77\quad\pm18$
Cow	—	—	2	$44\cdot7$	2	$24\cdot0$
Rabbit	2	$15\cdot2$	2	$5\cdot1$	2	$2\cdot9$
Guinea-pig	3	$16\cdot4\pm9\cdot1$	3	$13\cdot8\pm1\cdot5$	3	$13\cdot1\pm0\cdot8$
Rat	—	—	10	$29\cdot3\pm6\cdot8$	—	—

Mean values ± standard deviation. In cows the rennet bag has been used, in rats the glandular portion of the stomach after removing the rumen.

n = number of animals tested, but in cats, rats and guinea-pigs the organs of ten animals have been pooled.

(Lorenz, *et al.*, 1969, *Biochem. Pharmac.*, **18**, 2625.)

The values of Table 29 are also at variance with determinations by Kim and Glick (1968), who, employing the non-isotopic method, found no significant formation of histamine in any histological region of the corpus of the stomach in the dog or pig. The Munich workers believe that the findings of low gastric mucosal HFC by other workers is due to the catabolism of newly formed histamine during the incubation process. To obviate this loss, they added aminoguanidine, chlorpromazine and nicotinamide with the object of inhibiting, respectively, diamine oxidase, histamine methyltransferase and diphosphopyridine nucleotidase. In some species (monkey, cow, guinea-pig), Lorenz *et al.* found that even in the presence of these inhibitors, a loss of newly formed histamine still occurred between 10 and 30 minutes of incubation. In our isotopic determinations no such loss was noted. Moreover, it appears that the presence in the gastric mucosa of the three histamine catabolizing enzymes cited above has not been established with certainty. The discrepancies and uncertainties referred to here are likely to stimulate further studies of gastric mucosal HFC in various species, and may finally lead to an agreement on the technique to be applied for its proper determination.

The presence of histidine decarboxylase in the gastric mucosa of monkeys and man has been reported by Noll and Levine (1970), employing an improved $^{14}CO_2$ method. The bulk of the enzyme was found in the region of the fundus. Enzyme activity was constant over the course of at least 3 hr of incubation, and the reaction rate was reported as strictly proportional to the concentration of the enzyme.

Distribution of HFC in the stomach wall

The high histidine decarboxylase activity of the rat glandular mucosa (corpus of the stomach) was discovered by Schayer (1956*b*, 1957) who emphasized that the enzymic activity was far higher than that from any other mammalian organ then tested.

The distribution of HFC in the gastric mucosa in various species has been explored by Kahlson, Rosengren, Svahn and Thunberg (1964). Their main results are as follows. The mucosal HFC is much higher in the fed than in the fasting state. In the pyloric region, which lacks parietal cells, the mean HFC in fasted rats was only a small fraction of that found in the acid secreting mucosa. On feeding, HFC elevation occurred only in this latter region, not in the pyloric mucosa.

The vertical distribution of HFC was studied by these workers in four fresh specimens of human stomach wall in which mucosa, submucosa and muscularis-serosa had been separated from each other. The mean HFC values in the three layers were 0·051, 0·011 and 0·007 μg/g, respectively. These workers suggested that the HFC of the submucosa and the muscularis-serosa probably was associated with mast cells which in the human stomach wall are found principally in these layers (Räsänen, 1958). Similar determinations in dog, rat and frog mucosa showed that in these species all but a small fraction of HFC is in the glandular layer.

In this study (1964) the distribution of HFC within the mucosal layers proper was established in eight specimens of human mucosa. Serial horizontal sections of frozen tissue were collected for HFC determination and microscopic examinations. In four cases a correlation between HFC and estimated density of parietal cells was found in the samples. Among the remaining specimens, one showed the same association, though less clearly, and one from a patient with gastric carcinoma and achlorhydria, which lacked parietal cells, had the lowest HFC of all specimens examined. In this connexion the investigations by Feldberg and Harris (1953) and by Smith (1961) should be recalled, who established in dogs and humans a 'histamine profile' and found the highest concentration of histamine in the region of the parietal cells. Reference to a correspondence between mucosal HFC and histamine content in man and dog should not be taken to imply that a correspondence of this kind is universal among mammalian species. The situation existing in other species will be set out under the next heading.

Relation between histamine content, HFC and sensitivity to histamine

Pertinent figures obtained in our laboratory and by other workers are assembled in Table 28. A conspicuous feature, and functionally significant, is the coincidence of very high mucosal HFC and very low sensitivity of the acid secreting cells to injected histamine, as seen in the rat and mouse. By contrast, the mucosal histamine content appears to be of the same order of magnitude in all species studied, presumably because the histamine content of the secretorily resting gastric mucosa depends primarily on circumstances which determine the binding of histamine and to a less extent on the HFC (Kahlson, Rosengren & Thunberg, 1963). The strikingly high HFC in the fasting rat and mouse is presumably not only related to the low sensitivity to histamine, but also to the presence of inter-digestive acid secretion in these two species.

Characteristics of histidine decarboxylase of the gastric mucosa

Schayer (1957) established the following criteria for histidine decarboxylase of rat stomach. The enzyme (1) is soluble, (2) has a high affinity for histidine, (3) differs from rabbit kidney decarboxylase which has a low affinity for histidine, (4) is *in vitro* moderately inhibited by semicarbazide or hydroxylamine, (5) is stable when

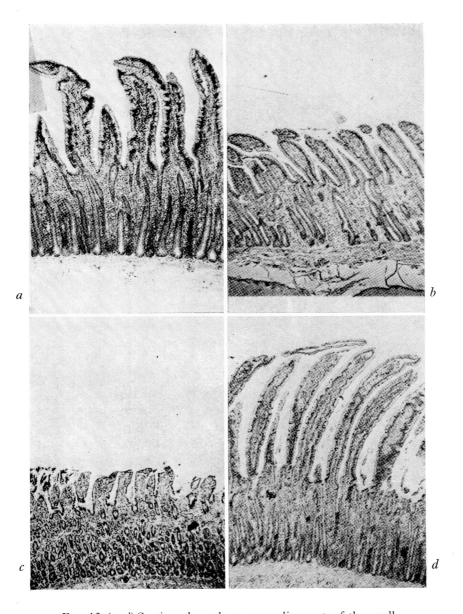

Fig. 13. (a–d) Sections through corresponding parts of the small intestine: (a) normal control, (b) hyp. ect., 20 days p. op., (c) adren. ect., 11 days p. op. Body wt. loss 14 per cent, (d) red. food during 11 days. Body wt. loss 12%. All sections are photographed with the same magnification. Note the reduction in villus length in (b) and (c) (Haeger et al., 1953b, Acta physiol. scand. **30**, Suppl. 111, 161).

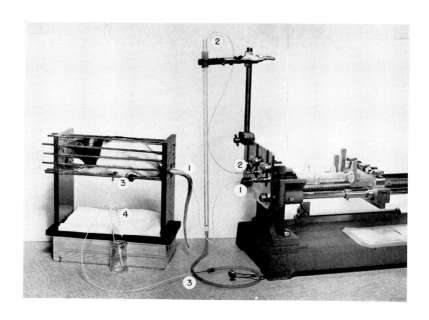

FIG. 24. Unanaesthetized rat in a restraining cage. Motor-driven infusion machine operating two syringes. 1. Catheter for intravenous infusion in the tail vein. 2–4. Arrangement for draining the pouch (Svensson, unpublished).

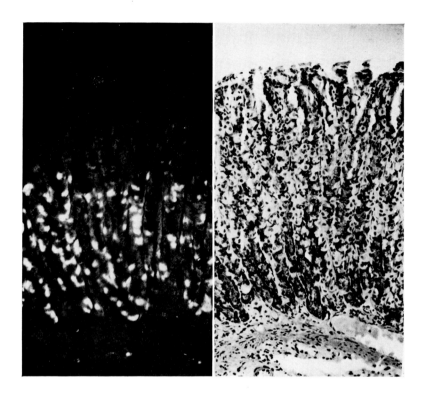

FIG. 39. Micrographs of rat gastric mucosa showing OPT induced fluorescence of histamine-containing cells (left part) and the same section subsequently stained (right part). Zeiss Standard Universal fluorescence microscope, UG 1/3 primary filter and secondary filter cutting off wavelengths shorter than 410 mμ. Stain: Mayer-Aurantia; parietal cells clear yellow, chief cells dark blue with cytoplasmic granules. × 80 (Thunberg, 1967, *Exptl. Cell Res.* **47,** 108).

stored in dry form at $-15\,^\circ$C, (6) has *in vitro*, like histidine decarboxylase of skin, brain and rabbit platelet, an optimal pH 7·2, as against 8·0 for rabbit kidney decarboxylase, (7) is strongly inhibited by benzene which activates the kidney enzyme, (8) and like histidine decarboxylase from other sources is affected by hormones.

Additional criteria came forth in a study by Kahlson, Rosengren and Thunberg (1963). From their work only three points will be mentioned here as the other findings have been previously discussed in the section on 'Inhibition of histamine formation'. (1) Gastric mucosal histidine decarboxylase is strongly inhibited by α-methylhistidine, but not by α-methyl-DOPA. (2) The dependence of the enzymic activity on pyridoxal phosphate was established quantitatively. (3) On dialysis the gastric mucosal histidine decarboxylase lost about 75% in activity, and this loss in activity could not be fully restored by the addition of pyridoxal phosphate. The pH for maximal enzymic activity *in vitro* was not determined in our study. Any such figure appears irrelevant with regard to physiological function as the decisive pH is that existing in the natural intracellular environment of the enzyme which at the present cannot be determined in the mucosa. In reports on the properties and kinetics of histidine decarboxylase it should be made clear that the observed characteristics pertain to extracts which may not necessarily be the same as for the enzyme *in vivo*. The extracted enzyme activity should not *a priori* be taken as identical with the properties of histidine decarboxylase in the intact cell.

Influence of feeding on gastric mucosal histamine formation

Feeding a fasting animal brings into play a sequence of secretory stimuli of which three have been recognized: increased vagal activity, release of gastrin and distension of the stomach wall. In due course a state of inhibited secretion ensues, in which the region of the antrum is strongly acid and 'enterogastrone' is believed to be released. In this state neither vagus stimulation, injection of gastrin or histamine, or any combination of these stimuli, is capable of exciting acid secretion in the innervated stomach at the high rates which occur with food in the stomach.

In unanaesthetized fasting rats provided with a Pavlov pouch, Svensson (1969) infused hog gastrin II in stepwise increasing doses until a maximal acid secretory reponse to gastrin was obtained, as seen in Fig. 23. At this stage, during continued gastrin infusion, the rat was allowed to eat pellets for 30 minutes. As a result the rate of acid secretion increased to about twice that obtained with gastrin alone. Noteworthy in this experiment is the relatively high rate of inter-digestive secretion existing before gastrin infusion, and the non-effect of gastrin on pepsin secretion. The arrangement used in gastric secretory studies in our laboratory is shown in Fig. 24 (between pp. 104–105).

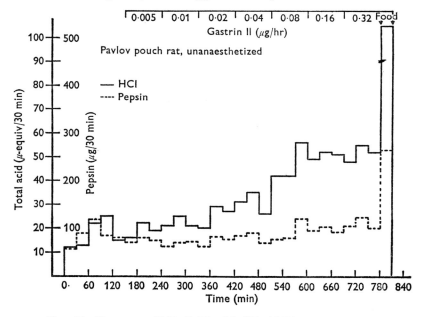

FIG. 23. (Svensson, 1969, *J. Physiol.* **200**, 116P).

The mechanism by which vagus stimulation and gastrin affect the acid secretory device is unknown. Notwithstanding, it has recently been disclosed that these two means of stimulation produce striking and identical changes in the histamine metabolism of the gastric mucosa. These changes will now be discussed.

At an early state of these experiments by Kahlson, Rosengren, Svahn and Thunberg (1964) in rats it was noted that the mucosal HFC varied widely in rats with free access to food. On closer examination it was found that withholding food for 24 hours resulted in a substantial lowering of mucosal HFC and that under fasting conditions the level varied only slightly in the test animals. Feeding, in the form of introducing 6 ml. dietary paste by stomach tube, was followed by an elevation of HFC. This amount of food distends the stomach wall considerably and corresponds approximately to the volume of food often found in the stomach in rats with free access to food. In this type of experiment, principal natural stimuli engaged in exciting gastric secretion co-operate in bringing about the elevation of HFC illustrated in Fig. 25. The HFC reaches a peak of about 7 times the fasting level at 3–4 hours after tube-feeding.

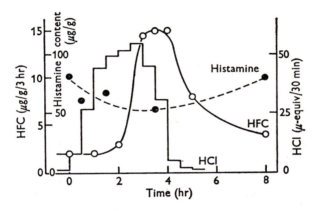

FIG. 25. Changes in HFC (open circles) and in histamine content (filled circles, broken line) of the gastric mucosa related to acid secretory response following re-feeding fasted rats. HFC and histamine content are means of figures obtained in a series of rats. The curve of HCl secretion is based on means from three experiments in a rat with a denervated pouch. Feeding was carried out at zero time (Kahlson, Rosengren, Svahn & Thunberg, 1964, *J. Physiol.* **174**, 400).

In the two other species studied in this respect similar changes occurred. The values obtained in fasting mice eating dietary paste are listed in Table 30. From this it is seen that the elevation of HFC of the stomach wall is of the same order as in rat gastric mucosa. In frogs fed on minced ox liver a similar elevation of stomach HFC occurred. In these two species in which only very small quantities of mucosa could be obtained by scraping, the whole wall of the glandular stomach was investigated, and it was established in special experiments that the elevation of HFC took place only in the mucosa.

Effect of injected gastrin on mucosal HFC

At the time (1964) when our first experiments were done, a dry powder gastrin preparation (Gregory & Tracy, 1961) was given to us by Professor Gregory. The histamine content of this preparation, as assayed in our laboratory, was found to be <1 ng/unit of gastrin. At later stages of our work a gastrin preparation of higher degree of purity, 'Gastrin II' (Gregory & Tracy, 1964), was also made available. We determined the stimulating potency of the two preparations and found that the hydrochloric acid secretory response in the denervated rat pouch of 4 units of the first preparation corresponded to 0·87 μg of Gastrin II.

In the first series of experiments 4 units of gastrin I was injected subcutaneously in rats fasted for 24 hours and the mucosal HFC was determined at various intervals afterwards. The maximum elevation produced by this dose of gastrin occurred earlier, at 1–2 hour, and was lower than that in response to feeding (Fig. 26).

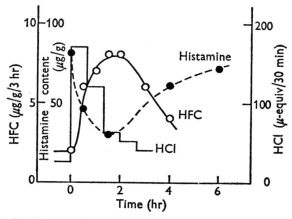

FIG. 26. Changes in HFC and in histamine content of gastric mucosa related to acid secretory response resulting from an injection of 4 u. of gastrin in fasted rats. Injection given at zero time. The curve of HCl secretion represents the mean of sixteen experiments in eight rats provided with a whole-stomach fistula (Kahlson, Rosengren, Svahn & Thunberg, 1964, *J. Physiol.* **174**, 400).

Next, rats were injected with increasing doses of gastrin I and the mucosal HFC was determined at $1\frac{1}{2}$ hr after the injection. As seen in Fig. 27, increasing the dose of gastrin above 10 units did not further elevate the HFC as determined $1\frac{1}{2}$ hr after the injection.

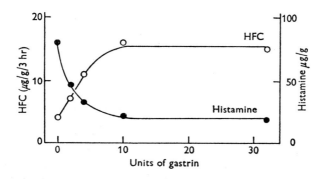

FIG. 27. Elevation of HFC and fall in histamine content of gastric mucosa in Sprague–Dawley rats at $1\frac{1}{2}$ hr after injection of various doses of gastrin. Each point on the curves is the mean of at least two determinations (Kahlson, Rosengren, Svahn & Thunberg, 1964, *J. Physiol.* **174**, 400).

Strikingly high elevations of HFC were achieved by dividing a single dose into small portions which were successively injected over a longer period (Fig. 28).

The magnitude of the elevation in HFC produced by repeated injections of small doses of gastrin for $3\frac{1}{2}$ hours equals, or even exceeds, the elevation seen after feeding. Experiments with a similar objective have been carried out in mice, cats and frogs. Single injections of gastrin in fasted mice produced an elevation of the whole stomach-wall HFC with peak values at 1 hour, whereas injecting histamine lowered the HFC (Table 30). The action of gastrin was also investigated in fasted, one-month-old kittens, litter mates being used as controls. In three kittens receiving repeated half-hourly injections of 4 units of gastrin and killed $5\frac{1}{2}$ hours after the first injection, the HFC was elevated about threefold as compared with three control animals (mean values: 0·43 and 0·16 µg/g, respectively). In two frogs, single injections of 0·8 units of gastrin elevated the whole stomach wall HFC about

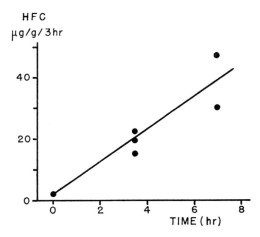

Fig. 28. HFC of gastric mucosa in fasted rats of the Physiological Institute, Lund, strain at $3\frac{1}{2}$ and 7 hr after repeated injections of gastrin (0·25 u. every half-hour). Control value at zero time is the mean for this strain (Kahlson, Rosengren, Svahn & Thunberg, 1964, *J. Physiol.* **174**, 400).

threefold at 1 hr as compared with four controls (mean values: 0·24 and 0·08 µg/g, respectively). Since, as already mentioned,

Table 30. HFC (µg/g) of stomach wall in mice. The mice were fasted for 48 hr before feeding or injections of gastrin and histamine. Voluntary feeding implies free access to food for 3 hr. Each figure is the mean of determinations in at least two mice

Voluntary feeding				
Time after onset of feeding (hr)	HFC	Time after injection (hr)	Gastrin (2 u.) HFC	Histamine (1 mg) HFC
0 (controls)	0·86	0 (controls)	0·86	0·86
3	2·4	$\frac{1}{2}$	1·4	0·70
6	5·8	1	2·8	0·48
9	0·8	2	2·0	0·42
12	0·72	$3\frac{1}{2}$	1·9	0·40
		$5\frac{1}{2}$	1·5	0·44
		8	1·2	0·85

(Kahlson, *et al.*, 1964, *J. Physiol.*, **174**, 400.)

elevation of HFC on injection of gastrin occurs only in the mucosa the HFC elevation of the mucosa proper must have been considerably larger than that determined in the whole stomach wall.

In human patients submitted to the augmented histamine test and insulin test, it was found that in both cases acid and pepsin secretion was inhibited to a similar degree by small doses of hexamethonium or atropine. From this, Clark, Curnow, Murray, Stephens and Wyllie (1964) concluded that the action of histamine in causing gastric secretion in man was not a specific one on the acid secreting cells, and that its action was not directly upon these cells but rather through nerves. They also considered it unlikely that histamine was the naturally occurring local chemostimulator of acid secretion in man. Bennett (1965), from a study of effects of gastrin on isolated rat smooth muscle preparations believed that gastrin acts on post-ganglionic parasympathetic nerves in the ileum, presumably by the release of acetylcholine, and he finally suggested that gastrin stimulates secretion by acting on nerves to release acetylcholine. This latter suggestion appears to find some support by observations to be discussed later.

Kahlson, Rosengren and Thunberg (1967) investigated the effect of gastrin injection on rat gastric mucosal HFC after blocking post-ganglionic cholinergic influences with Hoechst 9980, piperidino-ethyl-diphenyl-acetamide (for references to this compound see Emmelin & Henriksson, 1953). Pretreatment of rats with this compound did not abolish the elevation of HFC produced by gastrin injection (Fig. 29 and Table 31).

Until very recently there was no clear evidence to show that even in the inter-digestive state gastrin is continuously released from the antrum. Svensson (1970b) resected the antrum in rats provided with Heidenhain pouches and determined the mucosal HFC in the pouch. After antrectomy the mucosal HFC fell from 10·9 to 3·5 μg/g. Furthermore, following antrectomy there was a corresponding reduction of the inter-digestive acid secretion in the pouch.

Vagal activity elevates gastric mucosal HFC

The changes in gastric mucosal HFC have been investigated in rats and cats under the influence of vagal excitation produced by insulin, and in rats also after injection of 2-deoxy-D-glucose (2-DG), which latter compound is alleged to excite gastric secre-

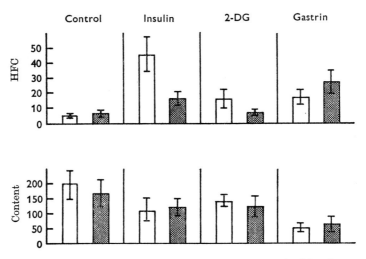

FIG. 29. HFC (µg/g) and histamine content (µg/g) of rat gastric mucosa, in controls and under the influence of insulin, 2-DG and gastrin. The open and the hatched columns stand for values obtained in animals injected with NaCl solution and Hoechst 9980, respectively, before the injection of NaCl solution (control), insulin, 2-DG or gastrin. Each column depicts the mean value and standard deviation of determinations in five rats (Kahlson *et al.*, 1967, *J. Physiol.* **190**, 455).

tion via the vagus nerves (Hirschowitz & Sachs, 1965; Colin-Jones & Himsworth, 1970). As in the experiments with gastrin, the cholinergic blocking compound Hoechst 9980 was employed in order to antagonize changes resulting from vagal excitation.

Insulin, 5 i.u./kg subcutaneously, elevated the mucosal HFC nearly tenfold, and this elevation was largely reduced by pretreatment with Hoechst 9980 (Fig. 29 and Table 31). Injection of 2-deoxy-D-glucose, 100 mg/kg subcutaneously, likewise induced a substantial elevation of mucosal HFC which was abolished by Hoechst 9980.

In cats, similar results were obtained. Eight young cats from two litters were examined. Each litter was divided into two groups, the one serving as controls was injected with NaCl solution, the other was injected with insulin. Results listed in Table 32 show that in cats insulin produced changes in HFC which are similar to those seen in rats. We (Kahlson, Rosengren & Thunberg, 1967)

Table 31. The upper half of the table details figures for mucosal HFC (µg/g), the lower half for histamine content (µg/g) in rats injected with saline or 9980 and under the influence of insulin, 2-DG and gastrin. The mean ±S.D. is also given

1 Control	2 9980	3 Insulin	4 Insulin + 9980	5 2-DG	6 2-DG + 9980	7 Gastrin	8 Gastrin + 9980
4·6	4·7	29·6	19·9	16·4	8·0	17·7	30·8
4·9	6·8	43·0	21·3	10·7	9·1	14·8	21·2
6·9	3·5	61·7	13·7	11·2	5·3	19·8	31·2
3·0	7·9	48·5	17·2	15·8	4·0	23·4	18·7
5·3	8·5	45·7	10·8	26·4	8·6	12·0	36·0
4·9 ± 1·36	6·3 ± 2·19	45·7 ± 11·57	16·6 ± 4·34	16·2 ± 6·30	7·0 ± 2·23	17·5 ± 4·38	27·6 ± 7·26
	1–2: diff. not signif.	1–3: $P < 0.001$	3–4: $P < 0.001$	1–5: $P < 0.01$	5–6: $P < 0.02$	1–7: $P < 0.001$	7–8: $P < 0.02$
225	211	160	104	149	114	46	96
228	191	51	167	119	137	73	33
240	144	134	107	119	72	54	49
143	190	73	127	152	122	44	63
142	92	127	95	169	173	54	82
196 ± 48·8	166 ± 47·9	109 ± 45·3	120 ± 28·8	142 ± 22·0	124 ± 36·6	54 ± 12·5	65 ± 25·2
	1–2: diff. not signif.	1–3: $P < 0.02$	3–4: diff. not signif.	1–5: diff. not signif. (P about 0·05)	5–6: diff. not signif.	1–7: $P < 0.001$	7–8: diff. not signif.

(Kahlson et al., 1967, J. Physiol. **190**, 455.)

did not explore the prerequisite for obtaining maximum elevation of mucosal HFC under the influence of insulin.

Table 32. HFC (ng/g) and histamine content (μg/g) of gastric mucosa in eight young cats injected with NaCl solution (controls) and insulin, respectively. Figures in parentheses refer to a cat with distended stomach

	HFC		Content	
	Controls	Insulin	Controls	Insulin
Litter 1 (860–930 g)	38, (80)	82, 92	32, (23)	23, 18
Litter 2 (550–800 g)	44, 70	82, 86	26, 27	31, 20

(Kahlson *et al.*, 1967, *J. Physiol.*, **190**, 455.)

INDIVIDUAL EFFECTS OF THE VAGUS AND ANTRUM ON MUCOSAL HFC

It is generally accepted that vagus stimulation causes release of gastrin. However, in spite of the large volume of publications on this topic and the incessant experimentation employing various methods there is still wide disagreement as to the magnitude of decrease in vagally induced acid secretion after removal of the antrum. In experiments with vagus stimulation by whatever means, e.g. insulin or 2-DG, the mucosal HFC is bound to rise under the combined effect of vagus stimulation and gastrin release. The effect of the gastrin component has been thoroughly established, but not the one exercised by the vagus.

Rosengren and Svensson (1969) demonstrated that in the rat the vagus and antral endogenous gastrin were each independently capable of sustaining and elevating gastric mucosal HFC. They investigated histidine decarboxylase activity in the parietal cell region in the following stomach preparations: gastric fistula, denervated Heidenhain pouch, antral resection with gastro-jejunostomy, gastrojejunostomy with exclusion of the duodenum and in the intact stomach. The determinations of mucosal HFC were made on fasting rats and on re-fed animals when the effect of feeding was studied.

In rats in which the antrum had been resected, Rosengren and Svensson (1969) investigated two problems: the consequence of eliminating endogenous gastrin and secondly, the effect of the vagus nerve *per se* on mucosal HFC. The controls comprised gastrojejunostomized rats with the antrum intact. These controls, for unknown reasons, displayed a higher mucosal HFC than that found in rats not subjected to any kind of surgery.

In the first place it was found that removal of the antrum was followed by a substantial fall in mucosal histidine decarboxylase activity from 8 μg/g in the controls to 2·3 μg/g (Table 33). This result again brings into prominence the major role of endogenous gastrin in sustaining mucosal histidine decarboxylase activity.

Table 33. HFC (μg/g) of gastric mucosa in antrectomized fasted rats after injection of saline, insulin, gastrin and feeding, and in fasted rats with a gastro-jejunostomy only. The means ±S.D. are also given. The significance of the differences in the means of different groups is indicated at the foot of each column

1 NaCl	2 Insulin	3 Feeding	4 Gastrin	5 Gastroje- junostomy only
2·0	2·3	6·0	6·7	9·4
0·8	7·1	12·0	5·4	7·7
1·9	3·3	3·3	5·3	10·0
3·0	5·4	3·9	3·9	9·0
2·5	3·3	4·1	6·2	9·0
3·8	5·1	5·0	8·4	5·6
				5·6
2·3 ± 1·0	4·4 ± 1·8	5·7 ± 3·2	5·9 ± 1·5	8·0 ± 1·8
	1—2: $P < 0.05$	1—3: $P < 0.05$	1—4: $P < 0.01$	1—5: $P < 0.01$

(Rosengren & Svensson, 1969, *J. Physiol.* **205**, 275.)

In the antrectomized rat the vagal influence can be investigated without interference by vagal gastrin release. An injection of insulin, 2·5 i.u./kg, evoked an increase of mucosal HFC to 4·4 μg/g (Group 2, Table 33). The possibility that larger doses of insulin would increase the enzymic activity still further was not investigated because higher doses of insulin than those employed are apt to inhibit hydrochloric acid secretion (Kim, Ridley & Tuegel, 1968).

The vagal discharge by insulin injection does not with certainty reproduce the vagal drive which occurs during feeding and, further, injected insulin may influence mucosal HFC by mechanisms other than vagal excitation. In order to test the effect of the natural vagal stimulation which is evoked on feeding, the fasting antrectomized rats were given food. Feeding for 3 hr elevated mucosal HFC from 2·3 μg/g to 5·7 (Table 33).

The finding that elimination of endogenous gastrin by antrectomy is followed by a fall in muscosal HFC made it *a priori* appear certain that injected gastrin would elevate mucosal HFC in the antrectomized stomach. This was confirmed by injecting 5 µg/kg of gastrin in antrectomized rats which were killed 2 hr afterwards. As seen in Table 33 mucosal HFC rose from 2·3 to a mean of 5·9 µg/g.

The mucosa of the intact stomach is under the influence of the vagi as well as of endogenous gastrin, whereas in the Heidenhain pouch vagal influence is absent and its mucosa is under the influence of endogenous gastrin alone. This difference in stimulatory influence is reflected in differences in HFC of the mucosa of the two stomach preparations (Table 34). The mean values obtained by Rosengren and Svensson (1969), 5·5 µg/g for the pouch as against 10·5 µg/g for the main stomach, indicate that in the inter-digestive state endogenous gastrin and the vagus both contribute in maintaining mucosal histidine decarboxylase activity, the vagal contribution being of considerable magnitude. In this connexion it should be mentioned that the Heidenhain pouch, even in the inter-digestive state, has been shown to secrete hydrochloric acid, although at a low rate (Kahlson *et al.* 1964; Lilja & Svensson, 1967). Inter-digestive secretion and mucosal HFC of the Heidenhain pouch are both substantially reduced after antrectomy (Svensson, 1970*b*). This would imply that in the fasting state endogenous gastrin is released in secretorily effective concentrations. The HFC which persists in the pouch after antrectomy is referred to as 'basal HFC' in this laboratory, believed to be related to metabolic processes in the mucosa.

Vagal influence appears not to be essential in gastrin-induced elevation of mucosal HFC. In the study of Rosengren and Svensson injection of gastrin elevated the HFC in the denervated as well as in the innervated mucosa, to 22·1 µg/g in the pouch, and to 26·8 µg/g in the main stomach (Table 34).

In the experiments on antrectomized rats a gastrojejunostomy was established to provide for natural feeding, and controls were likewise subjected to the operation of gastrojejunostomy. As a result, these controls displayed a substantially higher mucosal HFC than that found in rats not subjected to any kind of surgery or handling. In this group, incidentally, the HFC was found about as much elevated as in rats provided with a gastric fistula. The

Table 34. HFC (μg/g) in the mucosa of the pouch and the main stomach of twelve fasting Heidenhain pouch rats, of which six were injected with saline and six with gastrin. The means $\pm$S.D. are also given. The significance of the differences between the means of different groups is indicated at the foot of each column

Control		Gastrin	
1 Pouch	2 Main stomach	3 Pouch	4 Main stomach
7·8	17·0	17·0	24·0
9·1	12·0	36·0	20·0
3·7	4·9	8·8	34·0
5·8	15·0	49·0	55·0
2·2	7·8	12·0	11·0
4·5	6·5	10·0	27·0
5·5 $\pm$2·6	10·5 $\pm$4·9	22·1 $\pm$16·5	26·8 $\pm$15·8
	1—2: $P < 0.02$	1—3: $P < 0.02$	2—4: $P < 0.02$
			3—4: N.S.
			$P > 0.05$

(Rosengren & Svensson, 1969, *J. Physiol.* **205**, 275.)

reason for the high HFC in these two groups was not investigated by Rosengren and Svensson because this information was not essential for the conclusions they presented.

Experiments on antrectomized rats have been carried out by Johnson, Jones, Aures and Håkanson (1969) with the object of assessing the part played by endogenous gastrin in sustaining mucosal HFC. Their results agree with those of Rosengren and Svensson in the following respects: injection of penta-gastrin elevated mucosal HFC to the same extent in rats with and without the antrum; feeding elevated mucosal HFC even after antrectomy. The results disagree with Rosengren and Svensson in the failure to demonstrate any HFC in the antrectomized vagally innervated mucosa after fasting. As described above, vagal influence of itself, independent of the antrum, is capable of sustaining an elevation of HFC, and in the absence of both vagal influence and antrum a basal HFC persists (Svensson, 1970b). The failure to demonstrate mucosal HFC after antrectomy may be due to the $^{14}CO_2$-method employed by Johnson et al., the limitations of which are emphasized in Chapter 5.

Effect of distension of the stomach

Kahlson, Rosengren, Svahn and Thunberg (1964) introduced a small balloon, attached to a catheter, into the stomach of rats and filled it with 6 ml. of water. The rats were killed $3\frac{1}{2}$ hours later and the position of the balloon was ascertained. The balloon regularly distended the fundus and corpus region, but the antrum was only partially affected. The mucosal HFC in five rats subjected to distension was 3, 4, 4, 8 and 11 µg/g as compared with the mean of 2 µg/g in a group of nine controls.

In experiments on cats described earlier, a cat of litter 1 (figures in parentheses in Table 32) intended to serve as control, escaped from the cage during the period of fasting (Kahlson, Rosengren & Thunberg, 1967). This cat did not have access to food but the antral part of the stomach was found stuffed with wooden shavings and contained a highly acid juice; this was in contrast to the other controls in which the stomach was seemingly devoid of juice, and the surface of the mucosa covered with mucous of about neutral reaction. In this cat the mucosal HFC was 80 ng/g as compared with a mean of 38 in the litter mates.

Effect on HFC of stable choline esters

The influence of these parasympathomimetic compounds has been studied by Rosengren and Svensson (1969) because, as has been described, vagus excitation evokes a substantial acceleration of histamine formation in the gastric mucosa. Methacholine chloride, 200 µg/kg, and carbacholine chloride, 20 µg/kg, were injected subcutaneously in twenty-four rats, and groups of six were killed at 1 hour and at 3 hours after the injection. From Table 35 it is apparent that neither compound altered mucosal HFC significantly, although both, in the dosage employed, evoked profuse HCl secretion. The results would imply that these compounds activate the secretory device without a concomitant acceleration of histamine formation which ensues on vagal stimulation.

Apparently, these compounds act directly on the secreting cells, like histamine, the injection of which, as will be shown later, does not accelerate histamine formation. Suggestive of different modes of exciting the secretory cell is the finding of Svensson (1970*a*, *b*) that in the unanaesthetized rat the highest infusion rate of metha-

Table 35. HFC (μg/g) of gastric mucosa in fasting Sprague–Dawley rats after injection of choline esters. The means ±s.d. are also given. The figures obtained after injection of carbachol and methacholine do not differ significantly from those of the controls

Control	Carbachol		Methacholine	
	1 hr	3 hr	1 hr	3 hr
2·8	1·8	3·6	4·3	2·5
2·2	3·3	2·7	4·2	3·6
4·8	3·7	9·8	2·5	3·5
2·0	3·2	1·4	5·9	2·4
3·4	2·6	3·1	2·1	6·5
4·8	2·4	2·5	5·8	3·9
3·3 ±1·2	2·8 ±0·7	3·9 ±3·1	4·1 ±1·6	3·7 ±1·5

(Rosengren & Svensson, 1969, *J. Physiol.* **205**, 275.)

choline evoked a secretory rate of acid that was about twice as high as that obtainable with gastrin (Fig. 30).

After antrectomy Svensson (1970*b*) superimposed graded infusions of methacholine on background infusions of gastrin II at the dose levels 0·02 and 0·08 μg/hr in Heidenhain pouch rats. Background infusions of gastrin enhanced the responses to methacholine. With infusions of 0·08 μg/hr of gastrin the acid responses to low doses of methacholine (1 and 2 μg/hr) were not merely restored but even exceeded the response obtained before antrectomy (Fig. 30). Large doses of methacholine superimposed on either of the two doses of gastrin employed did not evoke acid responses of the magnitude obtained before antrectomy. Since in the rat Heidenhain pouch methacholine and gastrin accentuate the acid response when acting jointly (Svensson, 1970*a*), it would appear that the diminished response to methacholine after antrectomy is due to the absence of a continuous release of gastrin.

The dose–response curves of Fig. 30 may, perhaps, suggest that methacholine enhances the release of gastrin. However, this suggestion is disturbing, in that methacholine did not produce a measurable elevation of mucosal HFC. Further work is required in order to understand completely the circumstances governing the release of gastrin.

Biogenesis and physiology of histamine

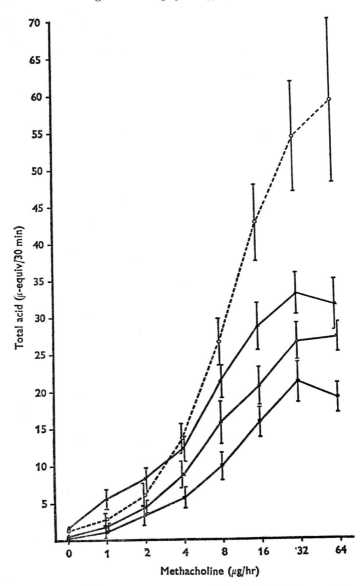

FIG. 30. Acid secretory responses of the Heidenhain pouch to intravenous infusion of methacholine in nine rats provided with a gastrojejunostomy before antrectomy (○ - - - ○), after antrect-

Circumstances restraining the elevation of mucosal HFC

Acid in the meal. Exposure of the antrum to high acidity is known to inhibit the gastrin release, and it thus appeared pertinent to investigate the effect of acid on the elevation of mucosal HFC. For this purpose a dietary paste was made up with HCl to a pH of 1–2 in experiments by Kahlson, Rosengren, Svahn and Thunberg (1964). In three rats tube-fed on 6 ml. of this acid paste the HFC was elevated only 1·5 times, compared with an elevation of 7 times in rats given the same paste at neutral pH. Another three rats were tube-fed on the acid paste and repeatedly injected with gastrin for 3½ hours as described earlier. The ensuing elevation of mucosal HFC was of the same magnitude as when gastrin was given to fasting rats. These results indicate that the effect of acid in the meal is to inhibit the release of gastrin rather than to interfere with its action on the HFC.

As already described, vagal stimulation evoked by insulin or 2-DG incites elevation of gastric mucosal HFC. It has also been explained that this elevation could be either due to a direct vagal effect on the histamine forming cells or to release of gastrin by the vagus, or to the combined effects of both.

At this point results obtained by Rosengren and Svensson (1969) in rats, will be described. Thirty-six rats provided with a gastric fistula were divided in six groups. Groups I and II were injected subcutaneously with 0·9% NaCl, groups III and IV were injected with 2·5 i.u./kg insulin, and groups V and VI with 5 μg/kg gastrin II. In the groups I, III and V the stomach was perfused with 0·9% NaCl and in the groups II, IV and VI with 0·1 M HCl. The perfusion commenced 1 hour before the injections and was discontinued 2 hours after injection of gastrin in the groups V and VI, and 3 hours after injections in the remaining groups. The timing was so set that the animals were sacrificed at the hour when the mucosal HFC, as known from earlier

---

Fig. 30 (*Contd.*)

omy (●—●), and after antrectomy in combination with a background infusion of gastrin in the doses of 0·02 μg/hr (△—△) and 0·08μg/hr (▲—▲). Each point represents the mean of determinations at the second and third 30-min period at each dose level with not less than one infusion in each rat. The vertical bars represent the s.e. of mean (Svensson, 1970*b*, *J. Physiol.* **207**, 699).

observations, was expected to be at peak level. From the results given in Table 36 it will be seen, in the first place, that gastrin, as well as insulin-induced hypoglycaemia, evoked elevations of mucosal HFC as described previously. Further, it is clear from Table 36 that acid in the stomach exerts a restraining influence on insulin-induced elevation of mucosal HFC. This restraining effect is presumably due to inhibition of gastrin release by the vagus when the stomach is acidified. This interpretation is strongly supported by the fact that acid in the stomach did not reduce the elevation of HFC evoked by injection of gastrin. It has been set out in a preceding paragraph that both endogenous gastrin and vagal influence contribute in maintaining histidine decarboxylase activity even in the fasting state. Acid in the stomach, assumed to inhibit release of endogenous gastrin, would therefore be expected to lower the mucosal HFC. This assumption was tested by the experiment in group II. With acid in the stomach the HFC fell from 8·9 to 3·4 μg/g within 4 hours. This result supports the conclusion that endogenous gastrin is of paramount importance in maintaining inter-digestive enzyme activity.

Table 36. HFC (μg/g) of gastric mucosa in fistula rats injected with saline, insulin or gastrin. In groups I, III and V the stomach was perfused with 0·9% NaCl, and in groups II, IV and VI with 0·1 M hydrochloric acid. The means ± s.d. are also given. The significance of the differences in the means of different groups is indicated at the foot of each column

I NaCl + NaCl	II NaCl + HCl	III Insulin + NaCl	IV Insulin + HCl	V Gastrin + NaCl	VI Gastrin + HCl
5·7	3·5	16·0	14·0	11·0	9·7
3·4	2·7	24·0	9·4	18·0	13·0
8·0	4·3	20·0	11·0	23·0	16·0
14·0	2·1	27·0	13·0	14·0	24·0
13·0	4·5	28·0	15·0	14·0	26·0
9·2	3·3	24·0	19·0	17·0	27·0
8·9 ± 4·1	3·4 ± 0·9	23·2 ± 4·5	13·6 ± 3·4	16·2 ± 4·2	19·3 ± 7·3
	I–II:	I–III:	II–IV:	I–V:	V–VI:
	$P < 0·01$	$P < 0·01$	$P < 0·01$	$P < 0·02$	N.S. $P > 0·05$
			III–IV:		
			$P < 0·01$		

(Rosengren & Svensson, 1969, *J. Physiol.* **205**, 275.)

Oil in the meal. Olive oil is alleged to stimulate the release of the inhibitory hormone enterogastrone, which is believed to operate by blocking the release of gastrin. Kahlson, Rosengren, Svahn and

Thunberg (1964) investigated whether the elevation of HFC produced by a meal would be altered by administration of olive oil. Olive oil (2 ml.) was given to rats by tube 1 hour before introducing 6 ml. of dietary paste which included another 3 ml. of oil. The mucosal HFC was determined 3 hours after introducing the paste. In four rats given oil the mean elevation of HFC was fourfold, whereas feeding five rats the dietary paste alone caused an elevation of sevenfold. When gastrin was injected repeatedly for $3\frac{1}{2}$ hours in rats given oil the resulting elevation of HFC was of the same magnitude as in control animals given gastrin only. These results indicate that the restraining influence of oil, like that of acid, is on the release and not on the action of gastrin.

HFC of the pyloric mucosa

In the rat the pyloric mucosa is devoid of parietal cells. In this region the mean HFC in four fasted rats was 0·07 μg/g as compared with the mean HFC in seven rats for the acid-secreting mucosa of 2·0 μg/g. In four rats killed $3\frac{1}{2}$ hours after tube-feeding, the HFC of the acid-secreting mucosa was elevated sevenfold, whereas the mean value for the pyloric mucosa, 0·08 μg/g, did not differ significantly from the fasting value (Kahlson, Rosengren, Svahn & Thunberg, 1964). With regard to ascribing increased histamine formation a role in exciting acid secretion, it is significant that the phenomenon of elevation of HFC by feeding or gastrin is strictly confined to the parietal-cell-containing region of the mucosa.

Acid secretion *per se* is not associated with elevated HFC

It has been shown on a foregoing page that injections of the stable choline esters carbachol and methacholine in doses which excite profuse acid secretion, do not alter the mucosal HFC detectably. When acid secretion is evoked by histamine injection the mucosal HFC is lowered. Kahlson, Rosengren, Svahn and Thunberg (1964) injected histamine in four rats in a dose corresponding to 2 mg of the base. Two rats were killed 1 hour after injection and two after 3 hours. The values for HFC at 1 hour were 0·8 and 2·8 μg/g, and at 3 hours 1·0 and 1·1 μg/g, all but one considerably below the mean fasting value of 2·0 μg/g. The same type of experiment was carried out in a large group of mice, each injected with 1 mg histamine and killed at various intervals. The results presented in Table 30 support the evidence obtained in rats of a lowering of mucosal HFC following injection of histamine. The character of this phenomenon, presumably a feed-back relation, is discussed in Chapter 8 and on the pages to follow.

Histamine excretion reflects changes in mucosal HFC

It has been explained in a previous paragraph that in the female rat changes in the rate of histamine formation are paralleled by changes in the urinary excretion of free histamine. In the rat a large fraction of the whole-body HFC exists in the glandular part of the stomach. It is *a priori* to be expected that substantial alterations in mucosal HFC should be paralleled by corresponding alterations in the urinary excretion of histamine. The actual situation is documented in Fig. 31. A period of fasting after voluntary feeding considerably reduced the histamine excretion as it has been shown to lower the mucosal HFC, whereas injection of gastrin resulted in increased excretion.

The fact that histamine, resulting from increased formation in the gastric mucosa, appears in the urine, indicates that the stomach of the rat lacks means for effective destruction of histamine formed therein. Claims to the contrary are based on determinations *in vitro* by methods of varying reliability (see Ivy & Bachrach, 1966, for references). The (whole) stomach of the guinea-pig, cat, rabbit and mouse, but not of the rat, yielded on extraction methylating enzyme activity, imidazole-N-methyl transferase. Brown, Tomchick and Axelrod (1959) suggest that the enzyme represents a defence against histamine for highly sensitive animals.

Excreted histamine, originating from elevated mucosal HFC, must have entered the blood stream and, on passage through the blood vessels, is bound to affect vascular tone. A plausible consequence of this circumstance will be discussed later.

Observations in the state of experimentally lowered HFC

The possibility of abolishing histamine formation throughout the body by inhibiting histidine decarboxylase activity has been discussed in some detail on preceeding pages. Investigating the secretory capability of a stomach completely devoid of HFC would obviously resolve the question whether the phenomenon of elevated mucosal HFC is an obligatory prerequisite in the process of exciting acid secretion. No means to eliminate mucosal HFC

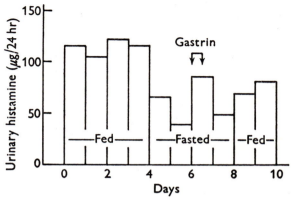

FIG. 31. Urinary excretion of free histamine in female rats
receiving aminoguanidine, 20 mg/kg daily. The values are the
means from experiments in three rats. 'Fed' stands for free
access to histamine-free diet. Whether fed or fasted, the rats had
water *ad libitum*. On the third day of fasting five injections each
of 4 u. of gastrin were given over a period of 15 hr (Kahlson,
Rosengren, Svahn & Thunberg, 1964, *J. Physiol.* **174**, 400).

completely has so far been discovered. Reduction by 90–95%
has been attained by means described in Chapter 8. In rats in
which gastric mucosal HFC had been reduced to about 5% of
normal the following observations were made (Kahlson, Rosengren, Svahn & Thunberg, 1964).

Secretory response. In normal rats fasted for 24 hours the mean
content of histamine in the corpus mucosa was 81 μg/g (Table 37,
which also shows the effect of feeding). In seven rats subjected to
injections of semicarbazide and fed on the pyridoxine-deficient
diet the corresponding value was 7·4 μg/g and the mucosal HFC
was reduced to about 5% of normal.

The secretory responses in rats in which the mucosal histamine
content and HFC had been thus reduced was investigated. Acid
output in response to injections of histamine and gastrin was
compared in the same rat before and during inhibition, by a test-
meal technique described by Thornton and Clifton (1959), and
adapted for use in our laboratory. Dose–response curves were
established for histamine and gastrin in individual rats and in the
actual experiments the two compounds were administered in

Table 37. Histamine content (μg/g) of gastric mucosa in rats fasted for 24–48 hr before re-feeding with 6 ml. dietary paste or injection of 4 u. of gastrin

Time after treatment (hr)	Re-fed	Mean	Gastrin injected	Mean
$\frac{1}{2}$	40, 42, 54, 67, 71, 75, 75	61	39, 43, 46, 54	46
$1\frac{1}{2}$	52, 54, 54, 60, 69, 76, 81, 93	67	23, 31, 39	31
$3\frac{1}{2}$	27, 31, 43, 44, 73, 75, 79	53	—	—
4	—	—	61	61
6	—	—	68, 75	72
8	59, 63, 96, 103	80	—	—

Controls: 51, 55, 58, 60, 62, 63, 71, 77, 78, 87, 100, 103, 110, 113, 123 (mean 81).

(Kahlson, Rosengren, Svahn & Thunberg, 1964, *J. Physiol.* **174**, 400.)

approximately equally active doses calculated to evoke substantial but submaximal secretory responses. A typical experiment is represented in Fig. 32. Secretory responses to histamine and

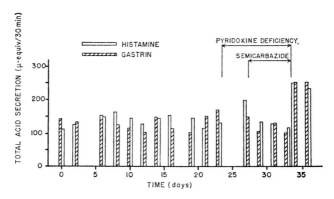

FIG. 32. Total acid secretion in a rat followed with the test-meal technique. Responses to injection of 4 u. of gastrin and 1 mg histamine (in terms of the base) were determined every 2–4 days before, during and after inhibition of histamine formation by semicarbazide (2 × 75 mg/kg daily) in a rat fed on a pyridoxine-deficient diet. Open columns, histamine-injected; cross-hatched columns, gastrin-injected (Kahlson, Rosengren, Svahn & Thunberg, 1964, *J. Physiol.* **174**, 400).

gastrin were determined every 2 or 3 days during 2–3 weeks while the animal was fed on an adequate diet. This was followed by 10 days of the inhibition treatment whereafter adequate diet was reinstituted. The response to gastrin and histamine, respectively, were determined on the same day with a 2-hour interval, the order of injections alternating as indicated in the figure. The animals were fasted for 15 hours before testing.

The principal finding in this experiment is that the secretory responses to injections of histamine and gastrin were approximately the same in the inhibited and in the normal state. That is to say, the amount of histamine, be it large or very small, contained in the gastric mucosa at any one moment, is not functionally related to the process of secretion and is not a factor determining the responsiveness of the parietal cells to histamine and gastrin. On discontinuing the inhibition treatment the responses to histamine and gastrin are larger, as seen in the four last columns of Fig. 32. This we explain as follows: in Chapter 8 we described the phenomenon of 'overshoot', which ensues on discontinuing the inhibition treatment, a phase during which histamine formation is increased throughout the body. Histamine thus generated, part of which is excreted, makes functional contact with the parietal cell whereby the inter-digestive secretion is presumably increased. The increased secretion adds to that evoked by injections of histamine and gastrin, and the combined secretory response becomes larger. Indeed, for increased histamine formation in the mucosa to be recognized as essential in the process of exciting acid secretion, then, in the inhibition experiment, the residual fraction of mucosal HFC should become elevated on injecting gastrin and on feeding.

HFC *can become elevated in the state of inhibition.* The seeming paradox that normal acid secretory responses to gastrin were obtained in rats in which the mucosal HFC and histamine content had been lowered to small fractions of normal has been resolved by the following experiments. In a group of nine rats subjected to inhibition treatment and fasting for 24 hours, the mucosal HFC in four was 0·06, 0·07, 0·11 and 0·12 μg/g (controls). Gastrin, 4 units, was injected in two rats whereupon at 1 hour the mucosal HFC was 0·30 and 0·35. In the remaining three rats the HFC-response to feeding was determined by tube-feeding 6 ml. pyridoxine-free nutrional paste. As a result, at 5 hours the HFC values were 0·12, 0·24 and 0·31 μg/g. That is to say, even in the inhibited state gastrin injection and feeding produced elevations of the residual HFC in nearly the same proportions as in the normal state. Two conclusions are to be drawn from these experiments. First, only a small fraction of the mucosal HFC suffices for effective association with the process of exciting acid secretion. Second, increased histamine formation in the mucosa is associated with this mechanism.

Similar experiments have been carried out by Levine and his colleagues who obtained results which are different from ours in two respects. In the first place these investigators searched for efficient *in vivo* inhibitors of histidine decarboxylase. Levine, Sato and Sjoerdsma (1965) have reported that in rats an alphahydrazine analog of histidine (α-HH) and 4-bromo-3-hydroxy benzyloxyamine (NSD-1055) lowered the histamine content of the stomach by 50% and the histamine excretion by 45%. In our laboratory, as described in Chapter 8, these two inhibitors were found to be only weak *in vivo* inhibitors of histamine formation, and attempts at explaining the differences in results between the two laboratories are set out in that chapter. Further, Levine (1965) reported that injecting these inhibitors in female rats, besides nearly depleting the stomach of histamine, decreased inter-digestive acid secretion and also significantly diminished the acid secretory response to insulin and gastrin. These results, which are entirely different from those obtained in our laboratory, cannot be accounted for by the assumption that the inhibitors may render the parietal cells insensitive to stimulatory influences, because the response to injected histamine was not diminished. There is, however, this difference in technique: we investigated secretory responses in intact rats, whereas Levine's studies were made on the pylorus-ligated stomach, as described by Shay, Sun and Gruenstein (1954), a rather mutilated preparation.

Thayer and Martin (1967) provided male rats with a stomach preparation *ad modum* Shay *et al.* and subjected them to treatment with NSD-1055 or a commercial pyridoxine-deficient diet (for at least 3 months). In both cases they reported a significant inhibition of the acid secretory response to injections of gastrin preparations, but no inhibition in response to insulin-induced vagal stimulation. The latter finding is contrary to the results reported by Levine (1965). The former authors give no information on the degree of inhibition obtained, except that no histamine forming activity could be found in 9 of 11 pyridoxal deficient animals.

Induction of changes in mucosal histamine content

Gastric mucosal histamine, even in the inter-digestive state, is formed at a considerable rate, leaves the stomach and is in part excreted in the urine. It should be emphasized that gastric histamine is in continuous movement within and across the mucosa, 'mobilized', as it were, and yet, in some species this histamine has no stimulating function, as evident by the absence of inter-digestive secretion in such species. Indeed, the turnover of histamine even in the unstimulated mucosa is very rapid. We have as yet no certain knowledge of the circumstances which govern the intragastric trails and functional associations of the histamine molecules formed in the gastric mucosa. Nevertheless, spectacular changes have been found to occur on feeding.

Feeding. The mucosal histamine content of rats tube-fed a dietary paste, after they had fasted for 24 hours, has been determined at various times after feeding. The means of individual determinations are plotted in Fig. 25. This figure shows that on feeding the mucosa is initially depleted of a small fraction of its histamine content and that restoration of the content takes about 8 hours. The initial lowering might have been completed within minutes; indeed, the time course of the very process of histamine depletion in the mucosa has not been determined and would be difficult to establish. Restoration of the depleted fraction is slow because the mucosa, like other rat tissues, does not take up and retain histamine other than that formed in the depleted cells and because the lifetime of histamine formed in the mucosa is brief.

Loss of histamine from the gastric mucosa on feeding has been demonstrated by Schayer and Ivy employing elaborate techniques. Their experiments are based on the following well established facts: injected ^{14}C-histidine is converted to ^{14}C-histamine in practically all tissues, including the gastric mucosa and the small intestine; ^{14}C-histamine thus formed leaves the tissues, enters the extracellular space and is in part excreted, after an interval depending on its tissue lifetime.

Besides a considerable degree of methylation, as emphasized in Chapter 7, the catabolism of histamine entering the blood and distributed throughout the body can be inhibited by aminoguanidine which inhibits histaminase which notably in the female rat is the principal histamine catabolizing enzyme. Schayer and Ivy

(1958) injected ^{14}C-histidine in re-fed starving male rats and noted that as a result of feeding the glandular stomach and the entire small intestine contained respectively 52% and 30% less ^{14}C-histamine than the non-fed controls similarly injected with ^{14}C-histidine. The lowering of the small intestine histamine content on feeding is noteworthy but it has so far not been studied by other workers. In another experiment, Schayer and Ivy (1957) injected ^{14}C-histidine into male rats first injected with aminoguanidine and determined the urinary excretion of ^{14}C-histamine, which they found to be increased on feeding as we found non-isotopically (Fig. 31). From their results they concluded that feeding caused a release of bound ^{14}C-histamine into the blood, that the resulting blood concentration of this histamine was capable of stimulating acid secretion, and that ^{14}C-histamine, thus released, is a blood-carried gastric secretory hormone in the rat. In male rats, as used in these experiments, methylation is the principal catabolizing mechanism and not oxidation by histaminase as in the female. The result of these early experiments would have been still more conspicuous, and more ^{14}C-histamine would have been excreted on feeding, if female rats had been used. The question of histamine as a blood-carried hormone will be discussed later.

Among the stimuli which in concert excite acid secretion on feeding, three have so far been recognized; vagal activity, gastrin and distension of the stomach. For each of them the influence on histamine formation in the gastric mucosa has been established, but less is known of their effects on the histamine contained in the mucosa.

Effect of injections of gastrin on mucosal histamine content. A first, and in retrospect unrewarding, approach to this problem was made by Emmelin and Kahlson (1944) who injected Komarov's gastrin extract in cats and from determinations of the histamine content of gastric juice believed they had demonstrated histamine release by gastrin. Next, Smith (1954, 1961) recorded histamine release from the stomach in cats on injecting a crude preparation of gastrin; this preparation also induced a more widespread release of histamine from isolated perfused skin flaps and skeletal muscle, and these effects he considered as perhaps experimental artefacts. This problem was reinvestigated by employing a less crude preparation of gastrin than that available to Smith in 1961. Haverback,

Tecimer, Dyce, Cohen, Stubrin and Santa Ana (1964) injected the equivalent of 10 g of whole hog pyloric mucosa subcutaneously in rats. They found the histamine content of the glandular stomach reduced by 15 to 30% at 1 hour of this injection and by 50% at 2 hours. Blair, Dutt, Harper and Lake (1961) injected gastrin extracts in anaesthetized cats, and determined plasma histamine levels by the method of Adam, Hardwick and Spencer (1957). From their results they concluded that gastrin is not a general histamine-releasing agent.

Purified gastrin I (Gregory & Tracy, 1961) of which 4 units in our laboratory caused near-maximum acid secretory responses in rats, as well as gastrin II (Gregory & Tracy, 1963) of which in our assay on the denervated rat pouch 0·87 µg corresponded to 4 units of gastrin I, were available in the studies by Kahlson, Rosengren, Svahn and Thunberg (1964). Injecting 4 units of gastrin in fasting rats caused a lowering of mucosal histamine content, the magnitude and time course of which are seen in Table 37 and Fig. 26. A reduction of about 50% was obtained with 2 units of gastrin and maximum lowering as well as maximum elevation of mucosal HFC was attained on injecting 10 units, as depicted in Fig. 27.

On comparing the changes in mucosal histamine, as illustrated in Fig. 25 and 26, it is easily seen that the reduction in content and elevation of HFC ensuing on feeding appear retarded against the corresponding changes caused by gastrin injection. A slow infusion of the same amount of gastrin, resembling the natural gastrin release on feeding, would produce changes of the retarded type.

Functions of histamine leaving the mucosa

It was stated earlier that histamine continuously escapes even from the unstimulated gastric mucosa. This steady stream of histamine is of considerable magnitude owing to a high rate of formation of histamine and its brief lifetime in the mucosa. The situation justifies, perhaps, an analogy with the packages of acetylcholine released from nerve terminals even at rest. Do stimuli that excite secretion hasten the stream of mucosal histamine as a nerve impulse hastens acetylcholine release? Presumably, yes, though we do not know with certainty to what extent, because at present no means can be envisaged to decide whether the histamine streaming on the one hand across the resting mucosa and on the other through the secreting mucosa has the same intracellular origin and moves along the same intramucosal pathways. Whatever function is finally assigned to the hastened stream of histamine within and from the mucosa on feeding, this change has been found to be coupled with a striking phenomenon, a concurrent elevation of gastric mucosal HFC.

Feed-back relation between histamine content and HFC in the mucosa

Until 1959 no means was known to inhibit histamine formation strongly *in vivo*. A specific inhibitor of histidine decarboxylase, α-methylhistidine, was recognized only in 1962 (Kahlson, Rosengren & Svensson, 1962). As a substitute for specific inhibition of this enzyme, rats were subjected to pyridoxine deficiency and injections of semicarbazide, as has been described in Chapter 8. On discontinuing this inhibitory treatment, the urinary excretion of histamine was much larger than before inhibition (Figs. 16 and 17). This phenomenon of 'overshoot', a rebound in histamine formation, we speculatively explained by the assumption that the reduction in tissue histamine content existing during the inhibitory treatment would adaptively enhance the synthesis of histidine decarboxylase, the unmasked activity of which emerged on reinstituting pyridoxine. At the time when the rebound phenomenon was observed, it had in fact been demonstrated that the depletion of rat skin histamine by 48/80 or polymyxin B concomitantly increased histidine decarboxylase activity in the skin (Schayer, Rothschild and Bizoni, 1959; Schayer & Ganley, 1959).

The speculation regarding a feed-back coupling was studied more closely in the gastric mucosa, a principal site of histamine formation. Figures 25 and 26 show that an initial lowering in histamine content of the mucosa triggers off, as it were, an increase in mucosal histidine decarboxylase activity. This coupling operates in both directions: injections of histamine lower mucosal HFC in rats and mice (Kahlson, Rosengren, Svahn & Thunberg, 1964), as already discussed in this chapter. It thus appears that in the intact tissues the actual concentration of the endproduct histamine controls the rate of the synthesis of the enzyme protein, accelerating or repressing the synthesis, respectively. The idea of a feed-back coupling is visualized in the way exemplified in a review article by Moyed and Umbarger (1962).

A similar feed-back mechanism has been suggested to operate in the control of the synthesis of 5-hydroxytryptamine. Contractor and Jeacock (1967), in an *in vitro* study on the guinea-pig kidney aromatic amino acid decarboxylase, discovered by Udenfriend, Clark and Titus (1952), found that 5-hydroxytryptamine inhibits this enzyme. In the preceding discussion of the properties of histidine decarboxylase we emphasized that these should be studied preferably in the natural environment existing in the living cell. In the test tube histamine does not repress the activity of histidine decarboxylase extracts. The formation of ^{14}C-histamine in mast-cell suspensions *in vitro* was not inhibited by preincubation with non-isotopic histamine (Schayer, 1956a). Levine and Watts (1966) added histamine (5×10^{-3} M) to a mixture of histidine decarboxylase extracted from foetal rat tissue and histidine and noted that the addition of histamine did not inhibit the formation of histamine in this mixture. On determining content and formation of histamine after stimulation of the submaxillary gland of dogs, Lorentz, Heitland, Werle, Schauer and Gastpar (1968) postulated a feed-back mechanism for the histamine–histidine decarboxylase system of the same kind as we found in the gastric mucosa. Injection of histamine was found by Kim and Glick (1968) to reduce the gastric mucosal HFC in the rat which to them was suggestive of a feed-back control of histamine synthesis.

Elevation of HFC is a prerequisite, not a consequence of excitation

In actively secreting glands it cannot always be ascertained whether an agent emerging on excitation is associated with the process of excitation or is a product of secretion. Kallikrein appearing in the saliva on stimulating the submaxillary gland in cats is a case in point (Beilenson, Schachter & Smaje, 1968). No uncertainty of this kind pertains to increased histamine formation in the gastric mucosa. Acid secretion *per se* is not accompanied by increased histamine formation in the gastric mucosa. If acid secretion is excited by injecting histamine, mucosal HFC is not elevated, but rather decreased, as discussed before. Elevation of mucosal HFC occurs on activating links in the excitation-conveying arrangement which are located proximal to the parietal cell. For this reason, clinical tests of gastric HCl secretion by injecting histamine will inform merely of the responsiveness of the parietal cells to the terminal stimulant histamine, but such tests will tell nothing of the functional state of the essential proximal components of the excitation conveying arrangement, where, in our belief, gastrin acts. For this reason, clinical tests for HCl secretion should be done by injecting gastrin (or pentagastrin), quite apart from the undesired side effects of a histamine injection.

Histamine as a blood-carried gastric hormone

The much disputed problem regarding the regular occurrence of histamine in blood plasma should now be considered as resolved. Histamine must be present in the plasma all the time, as evident by the steady excretion of histamine and its metabolites. Ever since Adam (1950), and Roberts and Adam (1950) emphasized that the output of histamine in the urine may serve as an indicator of changes in the plasma histamine level, the occurrence of histamine and its metabolites in the urine has been extensively studied and firmly established.

On taking food, and during the period of digestion in the stomach, the mucosal HFC is elevated and increasing amounts of histamine of gastric origin enter the blood stream. What are the consequences of the concurring elevation of the plasma histamine level? Schayer and Ivy (1957), as cited above, suggested that this blood-carried histamine excites acid secretion. This suggestion would seemingly find support by the view expressed by Hanson, Grossman and Ivy (1948), and by Öbrink (1948), that there appears to be a very low or no threshold for the stimulation of gastric secretion by histamine. On these premises histamine would be accorded the status of a blood-carried hormone. However, this conclusion appears less certain in the light of observations by other workers. Early work by Emmelin, Kahlson and Wicksell (1941) in dogs provided with a Pavlov pouch and given intravenous infusions of histamine showed that 0·5 µg per kg per minute was active, whereas 0·2 µg was inactive, in exciting acid secretion. They also noted, in cats and dogs, that increases in plasma histamine resulting from infusions which were too small to be detected by extraction and biological assay, produced vasodilatation, bronchoconstriction, increased motor activity of the bladder and secretion of gastric juice. Similar experiments were carried out in man by Adam, Card, Riddell, Roberts and Strong (1954). They infused histamine at a rate that stimulated acid secretion. Concomitantly there was a rise in the excretion of free histamine in the urine, but there was no detectable increase in the histamine content of plasma obtained from the antecubital vein during the course of infusions which stimulated gastric acid secretion. These authors restate the view that free histamine normally excreted in the urine is derived from the plasma. Lindell and Westling (1962) infused ^{14}C-histamine intravenously in man at a rate of 0·1 to 0·3 µg per kg/min which produced a distinct increase in pulse rate and facial flushing; the resulting increase in plasma histamine, determined isotopically, was too low to be detected by non-isotopic methods. These and similar results should be taken to indicate that the absence of non-isotopically detectable increase in plasma histamine does not rule out its active participation in the control of physiological functions.

The amounts of histamine which escape into the circulation during digestion in the stomach are too small, in unit time, to attain a magnitude of plasma level required for effective and sustained excitation of acid secretion. This conclusion appears justifiable by available information obtained from determinations of the elevation of mucosal HFC on feeding or under the influence of the known individual excitatory components which operate on feeding. Notwithstanding, the secretorily subthreshold level of plasma histamine may possibly provide a stimulatory contribution by adding to, and thus enhancing, the secretory effect of the other stimuli for acid secretion, i.e. vagal activity, gastrin and distension. The magnitude of the contribution by plasma histamine, if any, would be difficult to assess and it appears a matter of sheer speculation whether the dose–response relation to small infusions of histamine has any bearing on the problem (solid line of Fig. 36). In the rat the amount of histamine infused has to be increased ten times in order to obtain a significant increase in acid secretion, whereas in dogs doubling the amount of histamine infused approximately doubled the rate of secretion (Code, Blackburn, Livermore & Ratke, 1949). This wide discrepancy between the slope of the dose–response curves should, perhaps, not be taken to indicate a species difference in the secretory effectiveness of endogenous mucosal histamine.

Elevation of mucosal HFC and time course of acid secretion

Here, as in other gastro-intestinal secretory responses, it is not possible to determine precisely the time lag between the commencement of the stimulus and the onset of the response. The measurable latency in response to feeding is long, in part because the increment in the stimuli involved is slow and progressive (Fig. 25). If the stimulus ascends steeply, as on injecting gastrin, the elevation of mucosal HFC is correspondingly steeper.

The HFC of the mucosa persists elevated and the histamine content remains lowered even after the rate of HCl secretion has come down to the inter-digestive level. Here, again, a situation exists in which histamine is formed and is leaving the stomach at considerable rates without stimulating acid secretion above the inter-digestive level. The stimulatory ineffectiveness of this histamine is the more surprising since the observations in the state of inhibited histamine formation disclosed that secretory responses can be elicited under conditions where mucosal HFC and histamine content have been reduced to a minimal fraction of normal. At first sight, the persistence of HFC at nearly peak value as well as maximum lowering of histamine content during a phase of steep fall in HCl secretion, may be accounted for by the feed-back relation, whereby the elevated HFC is bound to replenish the depleted histamine store. However, this reasoning does not explain the fact that although near-maximum changes in gastric histamine persist, these changes are inefficient in exciting HCl secretion. Conceivably, a shift in the functional status of gastric histamine has occurred. The maintained elevation of HFC has a time course associated with the processes of mucosal restoration and anabolism ensuing on gastric digestion. Apparently, the circumstances which govern mucosal HFC are complex. It remains to be discovered why in the post-secretion phase elevated mucosal HFC, even at peak values, is not effective in exciting acid secretion, and what the physiological demarcation is between the two kinds of elevated HFC which both result from increased histidine decarboxylase activity. The particular circumstances discussed in some detail have so far been explored in the rat. It should be recalled, however, that a distinct gastric mucosal HFC has been found in all species so far investigated, and that elevation of mucosal HFC on feeding or gastrin injection has been shown also in cats, mice and frogs.

The observations of an association between elevation of mucosal HFC and acid secretion have attracted the consideration of other investigators. Endeavours to confirm, or refute, our work have at times followed rather devious courses leading to equivocal results and conclusions. Nonetheless, the classical problem of histamine as related to acid secretion should be approached by all appropriate kinds of experiment. An indirect, pharmacological approach has recently been resorted to, employing injections of secretin which is known to inhibit HCl secretion in certain species. Caren, Aures and Johnson (1969) have reported that injection of secretin in rats 'did not block the activation by pentagastrin of histidine decarboxylase in the gastric mucosa', and that pentagastrin acts 'directly to activate histidine decarboxylase'. Johnson and Tumpson (1970), on injecting secretin in rats, became impressed by their observation that in rats with a gastric fistula, injected secretin had no effect on maximally histamine-stimulated acid secretion. On this ground these authors were 'left with the conclusion that one or both of the assumptions in Kahlson's hypothesis are incorrect'. Neither of the two groups investigated alterations of mucosal histamine content, and the enzymic activity was determined with the $^{14}CO_2$ method. Briefly, our pertinent observations are these: feeding and gastrin evoke an elevation of mucosal HFC which is preceded by a fall in mucosal histamine content which (presumably by a feed-back coupling) initiates the increased histamine formation which latter is associated with increased HCl secretion (secretory device, Fig. 37). In our laboratory we encountered elevated HFC during the phase of declining HCl secretion in 1964 (Kahlson, Rosengren, Svahn & Thunberg) (Fig. 25 and 26), and in 1968 (Kahlson, Rosengren & Svensson) in a study of the gastric secretion antagonist SC-15396 which we found to inhibit secretion and elevate mucosal HFC. At this point we re-emphasize the complexity of the status of gastric mucosal histamine and, awaiting fresh experiments, remain rather reluctant to make categorical statements on this topic.

Maximal acid secretory response to histamine and gastrin

This subject will be discussed briefly, with the object of showing that significant information regarding the role and mode of action of histamine as a natural stimulant of acid secretion is unlikely to come forth from comparisons of maximal secretory responses. The results on record are equivocal depending on techniques and species employed. In dogs with a denervated fundic pouch Grossman (1961) recorded that the gastrin-induced maximal rate of acid secretion varied from 12 to 35% of the maximum achieved with histamine stimulation under similar test conditions. By contrast, in rats with a gastric fistula, Adashek and Grossman (1963) found that the highest rate of secretion attained with gastrin was more than twice the maximal rate in response to histamine, and they consider this difficult to reconcile with the hypothesis that histamine is the final common mediator of excitation. In man, Makhlouf, McManus and Card (1964, 1965) noted that gastrin II was significantly more effective than histamine in evoking acid secretory responses. In a study of seemingly 'peak secretory rates' in man, Jepson, Duthie, Fawcett, Gumpert, Johnston, Lari and Wormsley (1968) found that the gastric acid outputs elicited by injections of gastrin I, pentagastrin, tetragastrin, histamine and pentagastrin snuff were very similar.

Investigators who, like Adashek and Grossman (1963) and Kahlson, Rosengren, Svahn and Thunberg (1964), obtained a higher maximum secretory response with gastrin than with histamine may find this difficult to reconcile with the view that histamine is a natural stimulant for acid secretion. However, this reasoning is not helpful, nor are comparisons of maximal secretory responses, for the following reasons. Injected histamine is a poor substitute for histamine formed locally where it is believed to act, as would, by analogy, injected acetylcholine be less active than acetylcholine formed and acting at a synapse. Histamine in larger doses releases adrenaline from the adrenal medulla, an agent which is alleged to inhibit acid secretion. Injected histamine and released adrenaline may in addition interfere unfavourably with the blood supply of the stomach.

Elevated HFC *is likely to increase gastric blood flow.* Owing to technical difficulties it has not been possible to measure the fraction of the metabolism of the gastric mucosa which is not involved in the process of acid secretion. Obviously, a resting metabolism must exist in the non-secreting mucosa (see Myren, Unhjem, Gjeruldsen & Semb, 1965, for references). It has been discussed previously that a 'basal HFC' is believed to be associated with this resting metabolism. In a study of the oxygen uptake by the gastric mucosa of the bullfrog, Forte and Davies (1964) found a consistent rise in respiration on stimulation with histamine. In order to supply oxygen, water and electrolytes, increased blood supply must accompany acid secretion. Histamine is very effective in increasing blood flow through the stomach. Cumming, Haigh, Harries and Nutt (1961) infused histamine acid phosphate (2·5–10 µg/kg per minute) in anaesthetized dogs and noted an increase in the rate of flow through the stomach during the phase of acid secretion. Gastric blood flow and oxygen consumption were measured in anaesthetized dogs by Cumming, Harries and Holton (1965) who gave gastrin II (0·13 µg/kg per minute) and histamine acid phosphate (0·6 µg/kg per minute) intravenously and found that both histamine and gastrin increased gastric blood flow and also stimulated oxidative metabolism when there was acid secretion. Gastric vasodilation with gastrin stimulation has been observed by Swan and Jacobson (1967). In unanaesthetized dogs with Heidenhain pouches these investigators found that during administration of progressively increasing doses of histamine the rise in secretory rate was paralleled by a progressive rise in mucosal blood flow. Intra-arterial administration of 20 µg/min of histamine to exteriorized segments of dog stomach increased blood flow at least 20 times, and the increase in flow preceded the onset of acid secretion (Moody, 1967). There is evidence for a shift in the distribution of blood from muscularis to mucosa during secretory stimulation (see Jacobson, 1967, 1968, for references). Fig. 33, from experiments by Jacobson, Swan and Grossman (1967), shows this shift.

The mucosal HFC is elevated on feeding. Increased amounts of histamine thus formed enter the mucosal blood vessels since the stomach lacks mechanism for effective catabolism of histamine. As a consequence gastric vasodilatation is apt to occur. The demonstration of gastric vasodilatation on stimulation with gastrin emphasizes the intervention of an intermediary agent, presumably histamine, because gastrin by itself does not cause vasodilatation.

In anaesthetized cats, employing the same amidopyridine clearance technique as the American investigators, Harper, Reed and Smy (1968) found that total gastric and mucosal flow increased linearly with increase in acid secretion, stimulated by histamine or gastrin, and that the increase in total flow was entirely due to the increase in mucosal flow. Over the range of secretion studied in

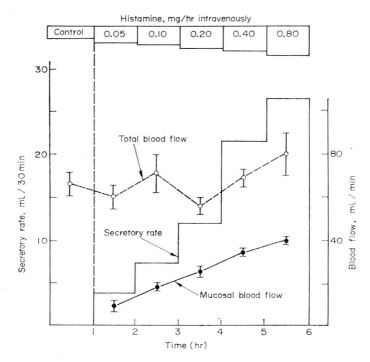

FIG. 33. Effects of graded doses of histamine on secretory rate, mucosal and total blood flow to the Heidenhain pouch (results from 4 dogs showing mean and standard error values). None of the total flow values differed significantly from control, whereas there was a significant increase in both secretory rate and mucosal blood flow ($P < 0.001$) with histamine (Jacobson *et al.*, 1967, *Gastroenterology* **52**, 414).

their experiments there was no difference between the effects of histamine and gastrin on mucosal flow. They believe that the fine adjustment of mucosal flow is determined by the activity of the oxyntic cells. Harper and his colleagues, strictly abstain from speculations. Yet, it appears reasonable to assume that increased mucosal histamine formation, associated with activity of the oxyntic cells, is a contributory factor in the adjustment of mucosal flow. A vascular function of histamine has been reported by Nylander and Olerud (1961). They demonstrated by a micro-angiographic procedure that a subcutaneous injection of histamine in rats evoked a diffuse filling of the mucosal arterioles and corresponding capillary net and infer that these changes are part of a shunting system of the gastric wall.

In sum: Histamine augments blood flow through the gastric mucosa. Gastrin-induced vasodilatation has been demonstrated with certainty only in the gastric mucosa. Gastrin appears void of inherent vasodilator property. Intravenous infusions of gastrin II, which stimulated gastric secretion, had no effect upon the rate of blood flow through human skin or voluntary muscle, as demonstrated by Ardill, Cumming, Fentem and Holton (1969), in Fig. 34.

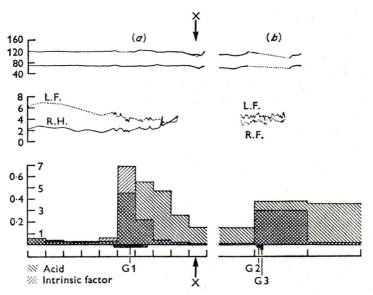

FIG. 34. The effects of infusions of gastrin II upon gastric secretion of acid and intrinsic factor and the rate of blood flow through hand (skin) and forearm (muscle).

Ordinates. Records from above downwards: arterial blood pressure in mm Hg; blood flow in ml./100 ml per min L.F. left forearm. R.H. right hand, R.F. right forearm; gastric secretion, left-hand scale, acid in m-equiv/min, right-hand scale, intrinsic factor in units $\times 10^2$/min.

Abscissae: Time marks every 15 min in both (a) and (b). (b) starts at × × in (a) and is continued with an expanded time scale in order to display the frequent blood flow readings. At G 1 intravenous infusion of gastrin II (50 µg in 28 min). At G 2, 2µg/min and at G 3 4µg/min of gastrin II were infused into the left brachial artery (Ardill *et al.*, 1969, *J. Physiol.* **203**, 57P).

It is difficult to visualize an arrangement which would prevent histamine formed in the mucosa at high rates from dilating the mucosal blood vessels, through which the amine does in fact pass.

The gastric-secretion antagonist SC-15396. An antigastrin which abolished gastrin-induced acid secretion without otherwise altering the responsiveness of the parietal cells to other stimuli would be helpful in elucidating the mode of action of gastrin. If increased mucosal histamine formation were an essential prerequisite for gastrin to induce acid secretion, a specific antigastrin compound should be expected to restrain gastrin-induced elevation of mucosal HFC.

The recently available compound 2-phenyl-2-(pyridyl)-thioacetamide (SC-15396) appeared to possess antigastrin activity in both rats and dogs (Cook & Bianchi, 1967). These authors also recorded that the compound inhibited experimentally induced gastric lesions in rats and duodenal ulceration in guinea-pigs (Bianchi & Cook, 1968).

This original work was carried further by others. Kahlson, Rosengren and Svensson (1968), whose report gives the pertinent references, investigated mainly the alleged specificity of SC-15396. First, these workers studied the effect of the compound on gastrin-induced gastric secretion. Gastrin II (2 μg/hr) was infused intravenously in Heidenhain pouch rats, and antigastrin, 1·5 mg per rat, dissolved in 50% polyethylene glycol, was injected subcutaneously. As seen in Fig. 35, following a transient peak, a constant rate of secretion was obtained about 2 hr after the onset of gastrin infusion. Antigastrin was injected during this plateau phase of secretion. Within 30 min of antigastrin injection, the HCl secretion was reduced, with an average inhibition of 70% with 1 mg of antigastrin. The duration of the inhibition resulting from one injection was about 3 hr. With 10 mg antigastrin per rat about 90% inhibition of acid secretion was obtained in rats bearing total stomach fistulae.

The gastrin-induced pepsin secretion was also reduced by antigastrin, but not so consistently as the inhibition of acid secretion. In five experiments the reduction was less than 10%, and in the remaining eight experiments it varied between 35 and 80%.

Next, inhibition of histamine-induced gastric secretion was investigated in Heidenhain pouch rats which had been injected

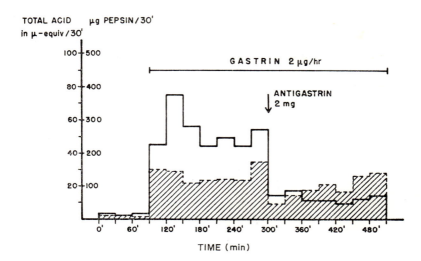

FIG. 35. Heidenhain-type pouch in an unanaesthetized rat. Ordinates indicate secretion of HCl (continuous line) and pepsin (cross hatched columns). Gastrin infused and antigastrin injected as indicated. Note low rate of spontaneous secretion, strong stimulation of both types of secretion by gastrin, and inhibition of both secretions by antigastrin (Kahlson, Rosengren & Svensson, 1968, *Brit. J. Pharmac.*, **33**, 493).

with mepyramine in order to obviate undesirable systemic circulatory effects of histamine. On successively increasing the infused dose of histamine, a level of secretion was achieved which approximately corresponded to the plateau maintained by gastrin infusion as in the experiment of Fig. 35. Injection of antigastrin, 1–2 mg per rat intravenously, reduced the rate of histamine-induced secretion of hydrochloric acid by about 75% (Fig. 36). Infusion of histamine at the rates employed in the experiment of Fig. 36 increased the secretion of pepsin in every rat. Injection of antigastrin reduced the pepsin secretion to about 40% of the rate prevailing before antigastrin.

The effect of antigastrin on the rate of histamine formation in the gastric mucosa was also investigated. Injection of various doses of antigastrin into rats during infusion of gastrin did not significantly influence the gastrin-induced elevation of mucosal HFC. To investigate further the effect of antigastrin on mucosal

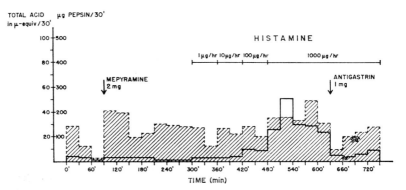

FIG. 36. Heidenhain-type pouch in an unanaesthetized rat. Ordinates and columns represent secretions as in Fig. 35. Note transient enhancement of spontaneous pepsin secretion by mepyramine. Histamine infusion and antigastrin injection as indicated. Strong and prolonged inhibition of histamine-induced HCl-secretion by antigastrin, strong but transient inhibition of pepsin secretion (Kahlson, Rosengren & Svensson, *Brit. J. Pharmac.*, 1968, **33**, 493).

HFC the secretory antagonist was administered before the injection of gastrin in twenty-four rats arranged in three groups. Group 1 was injected with antigastrin 5 mg per rat intravenously and 10 min later with 1 µg gastrin subcutaneously; group 2 was injected with the solvent used for antigastrin, followed by gastrin and group 3 with solvent and saline only. The animals were killed 2 hr after injection of gastrin. The gastric mucosal HFC mean values in the three groups were 16·5 µg/g, 11·3 µg/g and 4·7 µg/g, respectively. Antigastrin not merely failed to restrain gastrin-induced elevation of HFC but, rather, in itself increased mucosal histamine formation.

The workers in Lund arrived at the conclusion that SC-15396 acts by greatly reducing the sensitivity of the parietal and peptic cells to otherwise effective secretory stimulation, and not by directly interfering with the action of gastrin. This view is further substantiated by unpublished experiments by Svensson in which injection of 2 mg of antigastrin per rat strongly inhibited the secretion of acid and pepsin stimulated by intravenous infusion of mecholyl.

The search for a specific, non-toxic, antigastrin is likely to continue. Its success would provide, perhaps, the best available means of studying the mode of action of gastrin, and a place for antigastrin in therapeutics appears obvious.

THE HCl-SECRETORY DEVICE

The secretory device, as envisaged by Rosengren and Svensson (1969), is outlined in Fig. 37. It comprises the parietal cell proper, an adjacent structure in which histamine content and HFC are believed to reside, and nerve terminals conveying vagus activity. This outline does not take into account complex details of intramural structures related to the secretory device. This simplified outline appears useful in that it complies with experimental findings. The intractability of histamine as related to acid secretion is perhaps best met by provisional mental pictures which synthesize relevant observations.

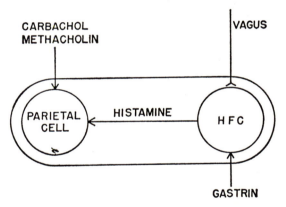

Fig. 37 (Rosengren & Svensson, 1969, *J. Physiol.* **205**, 275).

Localization of histamine and HFC *within the secretory device.* Until recently it has been tacitly assumed, without any kind of experimental proof, that mucosal histamine is contained in the parietal cell or in its nearest vicinity. However, current work indicates that this is not so. New information has been gained by a joint venture by the Departments of Histology and Physiology at Lund, employing histochemical techniques. The approach depends on the discovery by Shore, Burkhalter and Cohn (1959) that di-orthophthalaldehyde (OPT) forms a highly fluorescent product with histamine. This reaction has been employed by various workers for the chemical estimation of histamine in tissues (see Ehinger & Thunberg, 1967, for references). Ehinger and Thunberg (1967) elaborated a histochemical application of the reaction, referred to as the OPT procedure for histamine, intended for the cytological localization of histamine in tissues. A similar method was independently developed by Juhlin and Shelley (1966) and results obtained therewith have been mentioned in another chapter. The Ehinger–Thunberg technique appears to give a satisfactorily detailed picture of the distribution of histamine in tissues. Evidence that the visualized fluorescene is derived from histamine is based on the following observations: Fluorophores, apparently similar to those induced in mucosal sections, could be obtained from histamine in dry protein films; the fluorescent cells, histamine and HFC all occurred in one and the same part of the gastric glands; mast cells, known to be rich in histamine, appeared brightly fluorescent in all species investigated.

Thunberg (1967) applied this technique in a study on the localization of gastric mucosal histamine and HFC in female albino rats of the Sprague–Dawley strain. As a preliminary to the studies on fluorescence he examined sections 10 μ thick which were cut parallel to the mucosal surface in a cryostat at $-25\,°C$, about 40–60 sections being obtained in this way. Typical results are seen in the experiments illustrated in Fig. 38. Histamine and HFC were found to be equally distributed, and in all cases were confined to sections cut from the base of the gastric glands (terminology for base and neck region is in accordance with Stevens and Leblond, 1953). The parietal cells showed a maximal density close to the neck region which contained only little histamine and HFC. These results indicate the existence of a structure containing and forming histamine, located at the base of the gastric glands and apparently not cytologically related to the parietal cells. On stimulation of secretion changes in histamine content and HFC occurred, and they took place only in the base of the glands. In the fasting state much histamine was present and the basal HFC was low (Fig. 38a). On stimulation by injection of insulin, the histamine content was substantially reduced, and the HFC elevated (Fig. 38b). Feeding elevated HFC but reduced the histamine content only slightly (Fig. 38c). Gastrin II and 2-deoxy-D-glucose had effects

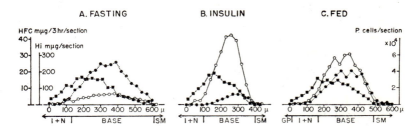

FIG. 38. Distribution of histamine (Hi), histamine forming capacity (HFC) and parietal (P) cells in the rat gastric mucosa, obtained by determinations on serial sections of the tissue. The rats were fasted for 15 hr and then injected with saline (*a*), insulin (*b*) or fed a meal (*c*) as specified in the text. The horizontal axis indicates the thickness of the sectioned part of the tissue from where the first complete section was obtained (usually 50–100 μ below the mucosal surface) and the histological structure seen in the sections (GP, gastric pits; I + N, isthmus and neck of the glands; BASE, base of the glands; SM, submucosa). The vertical axis indicates histamine ●——●, in mμg/section; HFC, ○——○, in mμg/3 hr per section; and parietal cells, ■——■, in cells/section (Thunberg, 1967, *Expl. Cell Res.* **47**, 108).

similar to those of insulin on mucosal HFC. Histamine, by contrast, in doses known to produce copious acid secretion, caused no observable changes. All these results are in complete agreement with earlier observations in our laboratory. The views regarding the localization of mucosal histamine and HFC are supported by Thunberg's observations with the fluorescence microscope. Fluorescent cells were abundantly present in the base of the gastric glands and occasionally in the submucosa, which latter cells could be identified as mast cells. The fluorescent cells at the base of the gland appeared to be neither parietal nor chief cells (Fig. 39, between pp. 104–105). This structure was also seen in control sections stained without preceding treatment with OPT, indicating that their staining properties do not result from this treatment.

Independent of Thunberg's work, and at the same time, Kim and Glick (1967) investigated the distribution of histamine and histidine decarboxylase in microtome sections of the body of the glandular stomach of the rat. The sections were 16 μ thick, 3 mm in diameter and 0·113 μl. in volume. Histidine decarboxylase was determined non-isotopically by measuring the amount of histamine formed from histidine in the reaction tube containing a tissue section. Histamine was determined by a fluorometric procedure. Enzyme activity was expressed as the difference in the values for preformed histamine and that obtained after the incubation. Fig. 40 from their report shows that histamine is formed and contained mainly in the base of the gland. The results of Kim and Glick agree with those of Thunberg.

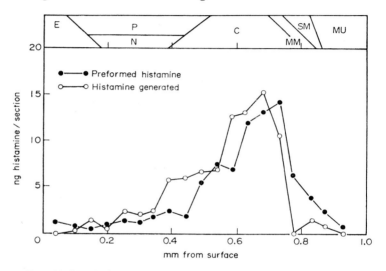

FIG. 40. Typical profiles of distribution of preformed histamine and histidine decarboxylase activity in the body of the stomach of the fed rat. Histologic zones indicated by letters: E, epithelial; P, parietal; C, chief; N, mucous neck cell; MM, muscularis mucosa; SM, submucosa; MU, muscle. Tissue section, 3 mm diameter, 16 μ thick (Kim & Glick, 1967, *J. Histochem. Cytochem.* **15**, 347).

Thunberg's discovery prompted Håkanson and Owman (1967) to examine the histamine forming structure of the mucosa histologically in some detail. These authors refer to the pertinent cells as 'enterochromaffin-like' because of seemingly morphological and histochemical similarities with enterochromaffin cells. The enterochromaffin-like cells are believed to form and store serotonin and dopamine under certain circumstances (Håkanson, Lilja & Owman, 1967), as judged by their fluorescence on applying the technique developed by Falck and Hillarp for visualizing catecholamines (Falck, Hillarp, Thieme & Torp, 1962).

HFC and secretorily ineffective gastrin derivatives

A participation of accelerated mucosal histamine formation, as an obligatory link, in the process of exciting HCl secretion, would be recognized as the more likely if it could be demonstrated that secretorily inactive derivatives of gastrin were also inactive in elevating mucosal HFC. A study with this object has been carried out by Rosengren and Svensson (1970).

The physiological effects of gastrin I and II are retained in derivatives with the C-terminal tetrapeptide sequence, Try. Met. Asp. Phe. NH$_2$ (Tracy & Gregory, 1964; Morley, Tracy & Gregory, 1965). Alterations in this terminal sequence impair the secretory potency of the compounds. The constitution of the peptide derivatives investigated by Rosengren and Svensson is given in Table 38 in which their structures are numbered according to Morley *et al.* (1965). Secretory responses were determined in rats provided with a Heidenhain pouch as described by Svensson (1970*a*). The rate of histamine formation was measured isotopically by our standard procedures. The compounds to be tested were injected subcutaneously, either once or repeatedly, at 30 min intervals. The animals were killed 30 min after the last injection.

The results obtained showed that among the compounds investigated a correspondence existed between the capacity to excite acid secretion and to enhance the rate of mucosal histamine formation. Hog gastrin II was the most active in exciting gastric secretion and accelerating mucosal HFC. Compounds 11 and 12, although secretorily less active than gastrin pentapeptide, did elevate mucosal HFC distinctly on repeated injections. Following a single injection of compound 16 a small acid response occurred. Repeated injections of this compound produced a significant elevation of mucosal histamine formation. The compounds 17, 24 and 25 produced a very small elevation of HFC, but did not evoke an appreciable secretory response (Table 38).

Behaviour of histamine within the secretory device. The nonsecreting gastric mucosa forms histamine continuously. This histamine, which results from the basal HFC, is continuously leaving the mucosa and replaced in a dynamic steady state fashion

Table 38. HFC (µg/g) in the gastric mucosa of rats after eight injections at 30 min intervals of control solution, hog gastrin II (10 µg/kg), gastrin pentapeptide (250 µg/kg) and analogues of gastrin tetrapeptide (250 µg/kg)

	Treatment	HFC (µg/g)	Mean ±s.d.
Control	SO_3H	2·5, 3·2, 3·2, 3·9, 6·4, 6·4	4·3 ±1·7
Hog gastrin II Peptavlon, ICI 50,123	⌐Glu. Gly. Pro. Try. Met. (Glu.)$_5$ Ala. Tyr. Gly. Try. Met. Asp. Phe. NH_2 21, 25, 31, 36, 48		32·2 ±10·5
(29)	BOC.β–Ala. Try. Met. Asp. Phe. NH_2	29, 31, 45, 48, 49, 51,	42·2 ±9·64
(11)	Z. Try. Met. Asp. Phe. NH_2	29, 33, 41, 48, 67,	43·6 ±15·0
(12)	Z. Try. Met. Asp. Phe. NH_2 —Phe. NH_2	25, 31, 35, 37, 39, 48,	35·8 ±7·75
(16)	Z. Try. Met. Asp.	8·8, 11, 12, 14, 19, 19,	14·0 ±4·54
(17)	Z. Try. Met. Glu. Phe. NH_2	3·9, 5·3, 6·0, 7·3, 8·2, 11,	7·0 ±2·49
(24)	BOC. Try. Met. Asp. Phe. OH	3·5, 3·8, 4·9, 6·6, 7·4, 7·7,	5·7 ±1·84
(25)	BOC.D.Try.D.Met.D.Asp.D.Phe.NH_2	5·3, 5·9, 6·4, 6·6, 6·8, 7·8,	6·5 ±0·85

BOC = Me_3 C. O. CO; Z = $PhCH_2$. O. CO.

(Rosengren & Svensson, 1970, *Br. J. Pharmac.* **38**, 473.)

governed by the lifetime of the histamine molecule formed, the feed-back relation and the rate of resynthesis. This fraction of histamine outwits effective stimulatory contact with the parietal cell, finds its way through the mucosa into the blood stream and appears in the urine as such or as its metabolites. The manner by which this is contrived is a matter of speculation. As a simple explanation it may be assumed that the basal HFC resides outside the secretory device. This assumption, however, is not entirely in agreement with the histochemical observations which suggest that in the gastric mucosa histamine content and HFC reside at one and the same site.

There is nothing bewildering in the finding that in the gastric mucosa histamine and HFC do not reside in the parietal cell, but in a different structure nearby. Moreover, this arrangement appears functionally entirely adequate. Histamine approaching the parietal cell from the outside causes the cell to secrete HCl as evident from the stimulatory efficiency of injected histamine. In this connexion a new circumstance has been disclosed which is bound to sharpen the picture. Within the gastric mucosa histamine formation takes place in a definite structure which in bulk represents only a tiny fraction of the whole mucosa, and is visualized as fluorescent spots. Even in the microsections, determinations of HFC in sections through the whole mucosa agree fairly well with our determinations on the whole mucosa. The HFC of the tiny histamine-forming structure must be huge as against the values calculated for the whole mucosa. As a consequence, histamine is made available in high concentrations near the parietal cell. Moreover, histamine is a freely diffusable agent and the mucosa lacks mechanisms for effective inactivation of histamine formed therein. It appears unreasonable to propose that parietal cells bathing in a stream of histamine naturally formed in their close vicinity would not be effectively stimulated by this amine.

What is the nature of the route along which histamine reaches the parietal cells in stimulatory effective concentrations? Transport and stimulation via the duct lumen proper should perhaps be ruled out, because histamine applied topically to the mucosa does not excite HCl-secretion (Hollander & Schapira, 1963). Histamine could diffuse upwards through the cell tube, through a number of non-parietal cells, though this appears unlikely. The most efficient transport would be provided by the net of capillaries which drains the gland. This route of histamine carriage appears the more likely since it is known that an increase in mucosal blood flow precedes the onset of secretion. An arrangement of this kind exists in the brain in which stimulatory agents are formed (released) in the region of the median eminence and carried in blood vessels to a target organ nearby, the anterior pituitary gland, which secretes in response to the blood-carried agents (see Chapter 13). Further, it appears conceivable that in the gastric gland, a kind of shunt mechanism operates by which histamine formed in the non-secreting state (basal HFC, interdigestive state in some species) is carried directly into the systemic circulation, thereby circumventing the parietal cells. Further experiments are likely to disclose the real situation.

Very little, if anything, is known of the mechanisms by which histamine excites the parietal cell to secrete HCl. Ash and Schild (1966), discussing receptors mediating some actions of histamine, support the view of a differentiation of histamine receptors into at least two classes among which so far only one type of receptor has been identified by its affinity for specific antihistamines. The intricate problem of the very mechanism of HCl formation has been reviewed by Davson (1970).

ACTION OF HORMONES ON HISTAMINE FORMATION IN THE GASTRIC MUCOSA

The relation between hormones and tissue HFC is discussed in Chapter 18. It is noteworthy that several hormones, on their natural release, or on injection, bring about profound alterations in histidine decarboxylase activity, i.e. changes in the rate of histamine formation in various tissues. The recognition of this hitherto neglected circumstance implies that histamine emerges as a member among the many agents which in physiological adaptations and responses are called upon to control tissue activities, and possibly to contribute in maintaining homeostasis. Among the tissue whose HFC is susceptible to influence by hormones, the gastric mucosa will be discussed in the following.

Corticosteroids. Having elaborated the isotope dilution procedure, which in an adapted form is used in our laboratory, Schayer (1956*b*) set out to investigate the rate of histamine formation in various rat tissues. His 3-page article contains, for the first time, the following information. The stomach is most potent in forming histamine, about 100 to 1000 times more active than the skin and the liver respectively. Cortisone increases histidine decarboxylase activity in the stomach. In the lung, the effect was opposite to that in the stomach, cortisone reducing the enzyme activity by about 50%; and prednisone nearly abolished lung histidine decarboxylase activity. Adrenalectomy reduced the enzyme activity in the stomach to about half and doubled the activity in the lung. Concurrent determinations of histamine content and histidine decarboxylase activity of tissues revealed to Schayer, and in 1958 to us, that there is no correlation between these values. It should be recalled that at that time much of the work on histamine was carried out in the belief that any conspicuous alteration in the histamine content of a tissue may provide a clue to the physiological significance of histamine. It was only later that this belief was dismissed.

The effect of adrenalectomy on histamine metabolism was first studied by Rose and Browne (1938) who reported that in rats following adrenalectomy the ability to inactivate injected histamine was markedly reduced, a change which could be restored to normal by injecting adrenocortical extracts (Rose, 1939). In mice, adrenalectomy did not alter the inactivation of ^{14}C-histamine injected in amounts which were much smaller than the histamine injections employed by Rose and Browne (Schayer, Smiley & Kennedy, 1952). The effects of adrenal insufficiency have been studied further by Bjurö and others. Their main findings are the following. In adrenalectomized rats maintained on 0·9% NaCl, injections of cortisone increased the urinary excretion of non-radioactive histamine. Similarly, cortisone increased the excretion of ^{14}C-histamine formed from injected ^{14}C-histidine. In both instances the increased histamine formation was presumed to take place mainly in the glandular part of the stomach. Determinations *in vitro* on the glandular stomach of adrenalectomized rats gave results which did not lend themselves to meaningful interpretation (Bjurö, 1963). In experiments by Bjurö, Westling and Wetterqvist (1964*b*) on hyperthyroid rats injections of cortisone increased the urinary excretion of non-radioactive histamine and of ^{14}C-histamine formed from injected ^{14}C-histidine. Their experiments will be further discussed.

Several investigators have reported on changes in tissue

F

histamine content following administration of corticosteroids, mostly omitting the gastric mucosa among the many tissues listed (see Hicks, 1965, for references). In a study of the effects of corticosteroids on the histamine content of rat tissues, Telford and West (1960) found that following nine injections of prednisolone or fludrocortisone for nine days, the histamine content of what they refer to as the pyloric stomach was about 3 times the control value (presumably the wall of the fundus was investigated). Heisler and Kovacs (1967) included the whole stomach among the tissues investigated. In pylorus ligated guinea-pigs they found that cortisone injections decreased the histamine content of the whole stomach and increased gastric secretory activity.

Thyroid hormones. On injecting thyroxine in rats, Parratt and West (1960) noted an increase in the urinary excretion of histamine, the increase depending on the number of thyroxine doses given. They attributed this increase to a diminished inactivation by histaminase of the histamine normally formed. The increase in histamine excretion was confirmed by Bjurö, Westling and Wetterqvist (1961) who, in addition, in rats given thyroxine, or liothyronine, demonstrated that the increase could be accounted for by an increase in histamine formation in the glandular part of the stomach to about twice the rate of the controls. These investigators (1964a) carried their study further and determined *in vitro* the formation of ^{14}C-histamine in eight different tissues of rats which had been injected with ^{14}C-histidine 10 min, 60 min and 22 hr before the determination of the ^{14}C-histamine formed. In the glandular mucosa of rats injected with ^{14}C-histidine only, the mean HFC at the three intervals were respectively 300, 640, 102, and when the rats had been treated with liothyronine the mean HFC were respectively 843, 1490, 104. The figures for HFC (i.e. the amount of histamine formed *in vitro*), stand for counts/min per gram of tissue.

This kind of experiment permits an approximate estimation of the turnover rate of histamine in various tissues. These investigators found a rapid turnover rate in the glandular stomach, kidney and liver, intermediate in lung tissue, slow in skin and small intestine. This estimate agrees with the determination of lifetime of histamine in various tissues, obtained by a different approach, and discussed in Chapter 8.

In sum: The gastric mucosal HFC is susceptible to alterations evoked by hormones which are not specifically stimulants or inhibitors of gastric secretion. The nature of this action is unknown and is, presumably, in some instances metabolic.

SECRETION AND MUCOSAL HFC IN THE STOMACH DURING PREGNANCY AND LACTATION

Gastric secretion during pregnancy in human and mammals has been the object of many investigations. Depending on the species and the methods employed, acid secretion has been reported as diminished, unchanged or even increased (see Crean, 1963, for references). In our laboratory the unanaesthetized rat has been found more useful than other species investigated in quantitative studies of gastric secretion under various circumstances. And the histidine decarboxylase activity of its gastric mucosa undergoes similar alterations under physiological circumstances as has been shown to occur in other species.

The pertinent changes during pregnancy have been extensively studied by Lilja and Svensson (1967) in rats provided with a whole stomach fistula or a denervated gastric pouch. In rats with a whole stomach fistula a steep increase in the basal secretion occurred from the first week of pregnancy onwards, and this increase reached a maximum during the third week. When pregnancy was followed by lactation the high inter-digestive acid secretion which existed during pregnancy persisted, and in some rats even higher values were seen (Fig. 41 and Table 39). The rate of the interdigestive secretion was 1·5–2 times the pre-pregnant values and 2–3 times during the third week and during lactation. The administration of aminoguanidine (AMG), which inhibits the histamine inactivating enzyme histaminase, did not appreciably affect the rate of secretion in the interdigestive phase. Injections of histamine or gastrin produced larger acid secretory responses during pregnancy and lactation. Following weaning the increased responsiveness gradually vanished.

In denervated pouches inter-digestive secretion was absent or very scanty, and did not change detectably during pregnancy and

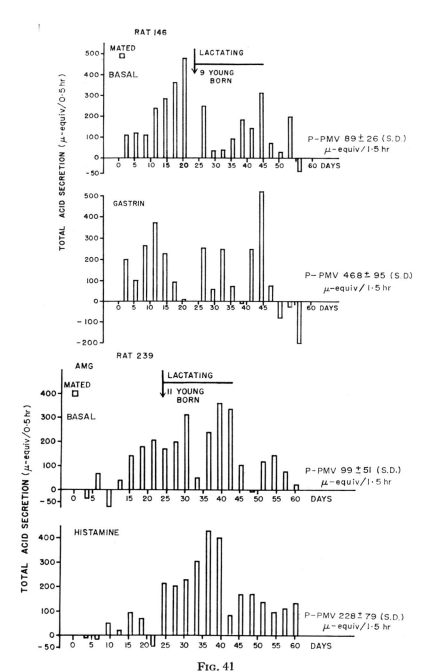

Fig. 41

Table 39. Basal, histamine (1 mg in terms of the base)—and gastrin (0·5 μg)—stimulated acid secretion during pregnancy and lactation in two groups of rats provided with whole stomach fistulas. The animals in the (*b*)-part of the table were under the influence of AMG. The figures with s.e. of the mean represent multiples of the P-PMV value. Pre-pregnant mean values (P-PMV) with s.e. of mean are expressed in μequiv/1·5 hr. The increase in basal secretion has been verified by applying the Student's *t*-test to the results

Period	Basal secretion	No. of rats	Histamine-stimulated secretion	No. of rats	Gastrin-stimulated secretion	No. of rats
			(*a*)			
P-PMV	169±24	6	286±23	5	346±32	1
Pregnancy						
1st week	1·2±0·11*	6	1·1±0·25	5	1·7	1
2nd	1·7±0·43†	6	1·5±0·37	5	1·3	1
3rd	2·8±0·37†	6	1·6±0·39	5	0·9	1
Lactation						
1st week	2·4±1·05†	6	1·8±0·49	5	1·3	1
2nd	2·2±0·71†	5	2·1±0·23	4	1·3	1
3rd	3·3±0·42†	4	2·1±0·59	3	1·8	1
After weaning						
1st week	1·9±0·30†	4	1·3±0·08	3	0·5	1
			(*b*)			
P-PMV	115±14	17	273±38	6	330±29	11
Pregnancy						
1st week	1·5±0·71†	17	1·2±0·29	6	1·1±0·42	11
2nd	1·9±0·60†	17	1·3±0·32	6	1·3±0·32	11
3rd	3·2±1·34†	17	1·3±0·44	6	1·2±0·35	11
Lactation						
1st week	2·7±1·44†	8	1·7±0·26	4	1·6±0·49	4
2nd	2·4±0·60†	5	2·8±0·30	3	1·8±0·43	3
3rd	3·5±0·78†	5	2·8±1·02	2	1·9±0·55	3
After weaning						
1st week	1·7±0·34†	4	1·53	1	1·3±0·30	3

* $P < 0.05$; † $P < 0.01$.
(Lilja & Svensson, 1967, *J. Physiol.* **190**, 261.)

lactation. A typical experiment presented in Fig. 42 shows that in the pouch the secretory response to histamine and gastrin is augmented, with peak values during the third week of pregnancy and during lactation.

FIG. 41. Basal and histamine (1 mg in terms of the base) or gastrin (0·5 μg) stimulated acid secretion during pregnancy and lactation in two rats each provided with a whole-stomach fistula. Rat 239 was under the influence of AMG. Rat 146 was not. The results are given as differences from a pre-pregnant mean value (P-PMV), obtained from five observations in rat 239 and seven in rat 146 (Lilja & Svensson, 1967, *J. Physiol.* **190**, 261).

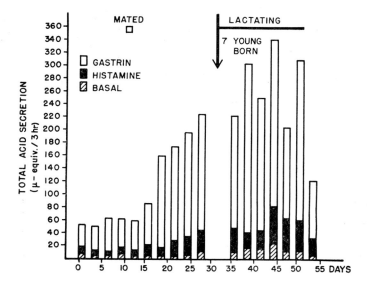

FIG. 42. Acid secretion before, during and after gestation in a rat provided with a denervated pouch. The columns represent a 3-hr collection, composed of 1 hr inter-digestive secretion (the cross-hatched part), followed by 1 hr of histamine (1 mg in terms of the base) stimulated secretion (the filled part), and finally 1 hr of gastrin (0·5 µg) stimulated secretion (the open part) (Lilja & Svensson, 1967, *J. Physiol.* **190**, 261).

Blood-carried histamine produced by the foetuses is not a main factor in evoking the increase in inter-digestive secretion in the whole stomach fistula rat, because AMG did neither augment this secretion nor promote secretion in the denervated pouch. The substantial amount of 2–6 mg histamine of foetal origin which in the rat enters the maternal blood stream in 24 hr (Kahlson, Rosengren & Westling, 1958) is seemingly subthreshold for acid secretion. Increased formation of histamine in the gastric mucosa proper could, however, account for the changes in secretion which occur during pregnancy and lactation. Lilja and Svensson (1967) investigated the mucosal HFC and, because mucosal HFC is elevated in fed animals, they made determinations in a fasting group and in a group with free access to food. Their results summarized in Table 40 show that in the fasted group the mucosal

HFC during the last days of pregnancy is about 2 times higher than in the non-pregnant controls, thereby providing for a higher concentration of histamine within the 'secretory device'. During lactation this increase persisted with mean values about 2–3 times those of the controls. In the group with free access to food the increase during the last days of pregnancy was about the same as in the fasted group. During lactation a few determinations were made on rats with free access to food and also here the increase was 2–3 times as against the controls.

Table 40. Mucosal HFC on different days of pregnancy and lactation

Day of pregnancy	HFC (µg/g) (fasting)	Mean	HFC (µg/g) (free access to food)	Mean
7	1·1, 1·6	1·4	—	—
13	2·3, 3·3, 4·9	3·5	6·2, 34·8	20·5
15	3·5, 4·0	3·8	15·0, 44·2	29·6
17	1·9, 2·1, 3·3, 4·3 5·6	3·4	43·1, 57·5	50·3
19	4·3, 4·8, 6·4	5·2	31·2, 58·2, 60·4	49·9
21	4·6, 9·1, 9·7	7·8	16·9, 42·8	29·9
Day of lactation				
1	3·0, 6·7	4·9	—	—
3	6·8, 15·2	11·0	—	—
7	3·5, 4·8, 6·5, 6·8 11·3, 11·7, 13·4	8·3	—	—
14	4·0, 5·7, 7·5 7·7, 12·4, 13·9	8·5	—	—
21	4·7, 6·5, 6·7, 8·8 10·4	7·4	—	—
Non-pregnant controls	1·7, 2·2, 3·8, 4·4 4·4, 4·5, 5·8	3·8	11·7, 13·3, 19·8 20·9, 31·0	19·3

(Lilja & Svensson, 1967, *J. Physiol.* **190**, 261.)

The increase in basal acid secretion during pregnancy and lactation coincides approximately with increased histamine formation in the mucosa. In addition, other changes may occur in the mucosa, a feature to be described below.

It should be mentioned that Code and Hallenbeck (1961), using Waton's non-isotopic method, have reported a three- to four-fold increase in the rate of histamine formation in the rat gastric mucosa from the 17th to the 21st day of pregnancy.

Protection against gastric ulceration during pregnancy and lactation

Investigators who work in fields where harvest ripens on time may wonder at the singlemindedness of those who devote so much effort in attempts to disclose the part played by histamine in the machinery of the body. On this point, Henry Dale's reflections, quoted in the Introduction to our book, should be recalled. Indeed, those dwelling upon the conundrums of the seemingly intractable histamine, may, in the course of their venture, find themselves embarked in some related, less unwieldy undertaking. An example of this will be given here.

During the last week of pregnancy the rat and the gastric mucosa is exposed to histamine, in part of foetal origin and, as has been discussed above, gastric acid secretion is concomitantly increased. Kahlson (1960) believed mistakenly that the foetal histamine provided the main effective stimulant for increased secretion, and that a non-pregnant rat subjected to long-lasting increased acid secretion would incur gastric ulcers. Although the idea prompting the study to be described turned out to be unfounded, nonetheless the pregnant rat proved to be ideal for studying the way in which the stomach is protected from ulceration.

In this study, Kahlson, Lilja and Svensson (1964) injected into non-pregnant rats the histamine liberator polymyxin B, which, as has been shown by Franco-Browder, Masson and Concoran (1959), regularly produces haemorrhagic erosions and necrosis in the glandular mucosa. However, in the later stages of pregnancy, injection of polymyxin B did not produce, or only rarely produced, gastric lesions, as seen in Table 41. In this table, one group of rats received aminoguanidine in order to retard inactivation of histamine liberated by polymyxin B. If the foetuses were removed on the 14th day of pregnancy, lesions appeared in nine out of ten rats. The presence of a single intact foetus protected the mother from gastric lesions.

The mechanism by which the foetuses protected the mother from gastric lesions was investigated by employing the technique of parabiosis. Female rats of the same litter, 5–6 weeks old, were parabiotically united, employing conventional surgery. At the age of 3 months, one partner was mated and became pregnant. On the 21st day of pregnancy polymyxin B was injected into both

Table 41. Lesions in glandular stomach after intraperitoneal injections
(2 mg/kg body-weight) of polymyxin B on various days of pregnancy

Day of pregnancy when polymyxin B was injected	No aminoguanidine No. of rats with lesions		Aminoguanidine (20 mg/kg daily) No. of rats with lesions	
..	5	(5)*	9	(9)
15	3	(3)	..	..
16	4	(4)	..	..
17	1	(1)	..	..
18	2	(3)	..	..
19	2	(3)	1	(2)
20	0	(4)	0	(3)
21	0	(3)	0	(3)
22	0	(2)	0	(3)

* Figures in parentheses represent total no. of rats examined.
(Kahlson, Lilja & Svensson, 1964, *Lancet*, **ii**, 1269.)

partners. In eight parabiotic couples, with one partner pregnant, polymyxin B failed to produce gastric lesions in any of the sixteen rats. In parabiotic couples, with no partner pregnant, polymyxin B produced lesions in the stomach of both. The parabiotic experiments demonstrated that the agent produced by the foetus was carried in the blood stream and, further, that it was present in sufficient concentration to protect even the non-pregnant partner against gastric lesions.

Evidence was obtained of a protective mechanism engendered by the act of suckling. In the experiments on two groups, summarized in Table 42, one group was suckling the young, in another

Table 42. Lesions in glandular stomach after injections of polymyxin B in lactating rats with suckling young, and in rats whose young had been removed immediately after parturition. (All rats received daily injections of aminoguanidine)

Day after parturition when polymyxin B was given	No. of rats with young with lesions	No. of rats without young with lesions
1	0 (5)*	1 (5)
2	0 (5)	2 (5)
3	1 (8)	4 (5)
4	1 (7)	4 (5)
7	2 (3)	..

* Figures in parentheses represent total no. of rats examined.
(Kahlson, Lilja & Svensson, 1964, *Lancet*, **ii**, 1269.)

group the young were taken away from the mother immediately after parturition. In this group of twenty non-lactating rats, polymyxin B produced gastric lesions in eleven, with a tendency for protection to decrease with the passage of time after parturition.

An investigation with a similar object has been carried out by Kelly and Robert (1969) who found that the incidence of peptic ulcer after treatment of rats with prednisolone is greatly reduced during the last third of pregnancy.

10

HISTAMINE AS A PHYSIOLOGICAL VASODILATOR CONTROLLING BLOOD FLOW

EVER since Carl Ludwig in Leipzig succeeded in recording continuously and in measuring changes in vascular tone and the rate of blood stream through tissues, the study of mechanisms believed to carry out the task of adapting the blood supply to the actual tissue requirements has retained its attraction. More is known of the nervous mechanisms involved than of the humoral factors. Only the latter will be discussed. Experiments aiming at revealing the role of humoral factors in the control of blood flow are guided mainly by three different assumptions.

I. Accumulated tissue metabolites. Nearly 100 years ago Roy and Graham Brown (1879), studying reactive hyperaemia in the frog's web, mooted the idea of 'a local mechanism independent of the centres in the medulla and spinal cord by which the degree of dilatation of the vessels is varied in accordance with the requirements of the tissues'. They stated: 'there will be a relative accumulation of the products of tissue change, which would probably, as in our experiments, stimulate the vessels to dilate'. In the century since, a large number of natural body constituents have been suggested as capable of meeting these requirements, and new aspirants are being continuously added to the list. No single factor, or combination of factors, have so far stood the test of experiment, as being capable of adapting flow to the need of blood. There is a standard review on the topic by Barcroft (1963).

II. Substance H of Lewis and histamine. This approach is somewhat different from the one discussed above. Thomas Lewis, and his followers, experimented and reasoned with the view that the dilator agent is released, and that the rate of release would correspond to the need of blood. This view of a dilator agent being released overlaps with Gaskell's (1880) idea that in the working muscle blood vessels are dilated by the CO_2 that is formed, and likewise by some agent released from the muscle fibres, possibly lactic acid.

The experiments and arguments of Lewis (1927), presented in *The Blood Vessels of the Human Skin and their Responses*, stand out as a forceful exposition of the H-substance as a physiological dilator agent. Lewis maintained on the basis of his experiments: 'With this series of facts before us it is difficult to refrain from stating without reserve the simple conclusion that the vasodilator substance considered and the H-substance are one and the same, and that this substance is histamine, free or held in loose combination. This conclusion would harmonize with the chief evidence at all points.'

The idea of released histamine as a factor in physiological vaso-dilatation was subsequently followed up experimentally by many investigators whose work is set forth in the Rocha e Silva (1966) Handbook. An obvious approach to the problem, as with histamine in gastric secretion, is to search for increased histamine in blood draining from the organ in which vasodilatation takes place. Extensive experiments, employing several tests for histamine in blood and plasma, were carried out in our laboratory. Emmelin, Kahlson and Wicksell (1941) obstructed the blood flow to the dog's ear, the dog's forelimb and the human arm, and they failed to find an increase in the histamine concentration of the venous blood drained from the ischaemic tissues, employing the Barsoum–Gaddum method for histamine extraction. In guinea-pigs and cats, using *in vivo* tests, namely the bronchiolar tone, the gastric secretion and the arterial blood pressure, they noted no evidence of excess histamine in blood when the abdominal aorta was un-clamped after having been obstructed below the renal arteries for 10–20 minutes. Tetanization of the dog's gastrocnemius, the dog's tongue and the muscles of the human forearm did not cause a detectable increase in the histamine concentration of the venous plasma from these muscles. As with similar experiments on hista-mine in gastric secretion, these results are equivocal owing to the relatively low sensitivity of the methods and tests available at that time.

Folkow, Haeger and Kahlson (1948) carried out experiments which appear worth mentioning for the following reasons. In the first place, a device was employed which made possible the continuous recording of the rate of blood flow on the smoked drum. This arrangement consisted of a Gaddum outflow recorder (Gaddum, 1929) and registering its readings by means of Fleisch's *'Ordinatenschreiber'* (Fleisch, 1930) adapted for use in our laboratory by Clementz and Rydberg, (1949). The very technique employed in this early work stands as the beginning of much of the work that followed in circulation research by several members of the group of physiologists of Lund. Incidentally we observed that after several hours of perfusion and slightly elevating the venous outflow pressure, the flow through the cat's hind limbs ceased completely although the blood pressure was still high. If at this stage the aorta was obstructed by a clamp, thus obviating the pulsatile pressure hammering on the constricted arteries, and the clamp was then removed, the arteries were found in a dilated state permitting a brief spell of ample flow (Fig. 43). This observation,

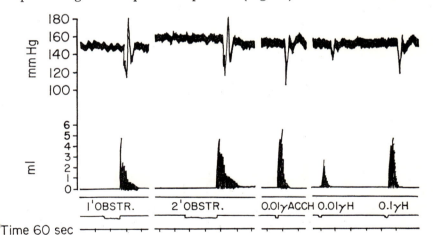

FIG. 43. Cat, 3 kg, under chloralose. Brachial artery blood pressure and blood flow in ml. through a hind leg. After a few hours of perfusion and slightly raising the venous pressure in the leg the blood flow is nil due to vasoconstriction. Obstructing the aorta for 1 min and 2 min, in the state of arrested blood flow, elicits vasodilatation as does intra-arterial injection of 0·01 µg acetylcholine, 0·01 and 0·1 µg histamine (Folkow *et al.*, 1948, *Acta physiol. scand.* **15**, 264).

using an advanced technique, reaffirmed Bayliss's discovery of the importance of the pulsatile pressure in sustaining arterial tone. Our observation was carried further in a detailed experimental analysis by Folkow (1949). The phenomenon seen in the two left sections of Fig. 43 should not be confused with reactive hyper-aemia; there was no flow to be obstructed, rather the pulsatile pressure was eliminated for respectively 1 and 2 min. Further, employing this flow-recording arrangement in twenty-four cats anaesthetized with chloralose, the dose of intra-arterially injected histamine required to produce a 25–50% increase for the first 30 sec in the rate of flow through the hind limb, varied in the range 0·003–0·24 µg/kg. Nevertheless, the magnitude and the duration of a standard reactive hyperaemia response was wholly independent of the individual sensitivity to injected histamine. Fig. 43 shows the vasodilator responses to injections of respectively 0·01 µg and 0·1 µg histamine. Pretreatment with the histamine antagonist Benadryl, reduced or abolished the vasodilator response to injected histamine in cats, and dogs under chloralose. Such pre-treatment did not alter the reactive hyperaemia response, as depicted in a typical experiment in a dog (Fig. 44). From these results it appears that blood-borne histamine is not a major factor in reactive hyperaemia.

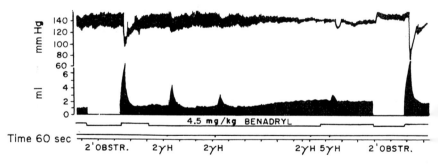

FIG 44. Dog, 6·5 kg under chloralose (Folkow *et al.*, 1948, *Acta physiol. scand.* **15**, 264).

Employing the same *Ordinatenschreiber*-Outflow technique, it was demonstrated that adenosine triphosphate (ATP) injected intra-arterially, nearly equals histamine in vasodilator potency, and that pretreatment with Benadryl or atropine, to abolish the effects of histamine and acetylcholine, did not diminish the vasodilator effect of ATP (Fig. 45). Regarding ATP, it would now, twenty years later, appear that the concentration of ATP in the venous effluent from exercising human forearm muscles is increased over the resting values, and that ATP was added to the blood in its passage through the exercised muscle (Forrester & Lind, 1969).

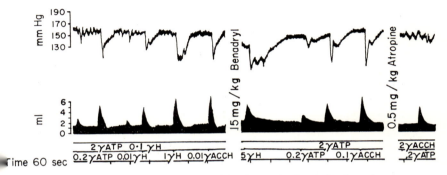

FIG. 45. Cat, 3·2 kg under chloralose. Action of adenosine triphosphate as compared with histamine and acetylcholine (Folkow *et al.*, 1948, *Acta physiol. scand.* **15**, 264).

In reactive hyperaemia, extravascular dilator agents, if any, are likely to accumulate primarily within the tissue spaces. For this reason it was considered essential to examine the lymph draining tissues the blood supply of which had been restricted. Lymph was collected from the hind legs of dogs, and to promote lymph flow legs were passively moved by a machine during the entire course of the experiment. The blood flow to the legs was arrested for 15 min followed by 5 min of free circulation. Ultrafiltrates or plain lymph emerging from the hind legs during reactive hyperaemia did not contain detectable amounts of histamine, but contained an agent which was conspicuously active in contracting the guinea-pig's ileum. The type of contraction resembled that of a 'slow-reacting' substance, the activity persisted after neoantergan and rendered the ileum preparation many-fold more sensitive to histamine. The activity occurred only occasionally in normal lymph. Incidentally it was found that the lymph, but not the blood plasma, contained histaminase (Carlsten, Kahlson & Wicksell, 1949*a, b*).

Similar experiments have been carried out by Edery and Lewis (1963). They collected lymph from the hind limbs of dogs before and after injuring the limb by various means including ischaemia for a period of 30 min. After injury, histamine appeared in the lymph and the kinin-forming activity of the lymph increased 2–9 times. Noteworthy in this connexion is that this result was seen also after ischaemia. Edery and Lewis concluded that the increase in kinin-forming activity after injury and ischaemia is the result of a specific activation of the kinin-forming system in the interstitial fluid, brought about by the release of histamine. From what we have observed in mice exposed to oxygen want, as described in Chapter 11, we would rather assume that increased histamine formation (not release) brought about the activation of the kinin-forming system. After ischaemia (30 min) the kinin-forming activity began to rise at 90 min. As far as humoral factors are concerned, some different factor must be brought in to bridge the lag between onset of vasodilatation and appearance of vasodilatory agents in the lymph. A possible mechanism is discussed below. Concluding this paragraph, it should be mentioned that very recently Ferreira, Corrado and Rocha e Silva (1969) have provided experimental support for Hilton's and Lewis's original theory (1955; 1957) relating kinins to functional vasodilatation in glandular organs and tissues. Further, Hilton and Torres (1970) and Gautvik, Hilton and Torres (1970) have brought forth experiments and arguments constituting a solid basis for the proposition that functional vasodilatation in glandular organs and possibly other tissues is mediated chiefly, if not entirely, by bradykinin.

III. Histamine formed in the arterial wall. Until recently, experiments on histamine as a physiological vasodilator were designed on the assumption that the amine was formed or, preferably, released in tissues outside the blood vessels in amounts prescribed by the actual need for blood. Thomas Lewis and Anrep are the principal adherents of this view. Their concept, unrewarding as it has appeared to later students of physiological vasodilator agents, has been revived by fresh views. It is mostly overlooked that arterial tissue contains histamine. Schmitterlöw (1948), in extracts from a variety of blood vessels of different species, found an agent which he considered to be identical with histamine; from the aorta of cattle he extracted about 10 µg/g histamine. Mongar and Schild (1952) extracted 17 µg/g histamine from the aorta of the guinea-pig. Riley and West (1953) recorded the presence of 0·7 µg/g histamine in the pig aorta in which mast cells were absent. Employing a histochemical procedure involving the formation of a fluorescent complex with orthophthaldialdehyde, Juhlin (1967) detected 1·4 µg/g histamine in the femoral artery of monkeys and 1·1 µg/g in rat cerebral areas rich in blood vessels, against un-detectable amounts in brain areas without vessels. With the staining method used histamine was seen along small arteries.

Schayer, like workers in our laboratory, maintains that tissues which contain histamine also form the amine, i.e. contain histidine decarboxylase. Pertinent to the topic under discussion is the discovery by Schayer that various circumstances are capable of increasing the tissue histidine decarboxylase activity. Schayer refers to this enzyme as inducible, and his views of 'induced histamine' are outlined in a review article (1966a). Schayer became impressed by the fact that circumstances and agents likely to induce histidine decarboxylase activity are accompanied by changes in tone and permeability of terminal blood vessels. The inducing factors are listed in a table in Schayer's review. These factors include local irritation, catecholamines, trauma, endotoxin, early and terminal phases of infection, Freund's adjuvant, or pertussis vaccine. The correlation between changes in histidine decarboxylase activity and circulatory changes is taken by Schayer as a basis for the hypothesis that induced histamine serves as a governor of the functional state of the terminal vessels, induced histamine being formed at a rate required to maintain homeo-stasis. Schayer believes that the inducible form of histidine de-carboxylase is located in or nearby the small blood vessels. The generalization which assigns to induced histamine the role of adapting the microcirculation to the actual requirements of the tissue is drawn from circumstantial rather than direct evidence. There is on record a single case in which HFC was measured in arterial tissue. In a study on anaphylaxis, Kahlson, Rosengren and Thunberg (1966) found that the guinea-pig aorta in the normal state had a HFC of 18 ng/g, a value which rose to 84, in the state of anaphylactic shock (Table 53). This single experiment indicates the possibility of 'inducing' histidine decarboxylase activity in arterial tissues. The gist of Schayer's concept lies in the im-portance assigned to the vasodilator agent produced within, not outside, the arterial tissue. Schayer recognizes that the intra-arterial histamine forming cells, in order to produce the required amount of the amine, must be liable to instruction conveyed by extracellular tissue changes. In this context he refers to an 'environment-responsive form of histidine-decarboxylase' (Schayer, 1968*b*).

The German surgeon August Bier, known to his contemporaries for the *Biersche Stauung* (applying reactive hyperaemia in limbs to promote blood flow and healing) spoke of the *'Blutgefühl'* of the tissues, their sense of blood. The meaning of the metaphysical *'Blutgefühl'* is that tissues in need call for more blood. In rational terms it may be asked: what is the nature of the call, and by what means is the call met? Conventional textbook accounts of locally produced vasodilator agents often refer to a wide list of propositions, e.g. from changes of pH to kinins. On this point it appears fitting to call attention to an investigation by the German physiologist Fleisch. He and Sibul (1933) investigated 63 at that time known intermediary products of metabolism for vasodilator activity on the perfused hind limb of the cat. The activity of each individual agent was very weak. However, Fleisch arrived at the conclusion that the vasodilator effect of these agents was additive, and that they, if acting in conjunction, would produce vasodilatation at a total concentration in the arterial blood of about 0·003 to 0·02 M. Granting that the metabolites in conjunction, or the increase in molarity by itself, bring about vasodilatation, the question remains unresolved whether these changes merely represent the call for more blood, i.e. the first link in a two-chain mechanism, or whether they constitute the very means to meet the call. The topic of humoral factors in the control of peripheral blood flow has been reviewed by Hilton (1962), Haddy and Scott (1968), Mellander and Johansson (1968), and Mellander (1970).

The astounding fact that after a century of elaborate investigation the nature of the physiological agent(s) controlling blood flow are still unknown would justify some doubt about the appropriateness of the ways by which this problem has so far been approached. Schayer, with a richness of arguments, has directed attention to a dilator agent formed within the blood vessels themselves. This approach, although technically difficult, may finally be found as not merely promising, but also very successful. Barcroft (1963), not specifically referring to Schayer's ideas, concludes the section of reactive hyperaemia with this remark: 'Can reactive hyperaemia be due to ischemia of the plain muscle of the arterial tree?'

From experiments on vasodilatation in the human forearm, Whelan (1956) remarked: 'If histamine is involved, it must be released in the cells on which it acts (the muscle cells of the blood vessels). . . .' Whelan finally concluded: 'This possibility of "intrinsic" histamine release was envisaged by Dale in 1948.' Schayer, like the present authors, would rather substitute increased histamine formation for histamine release. Indeed, chemical changes within the arterial wall, not merely outside them, may come out as crucial.

11

HISTAMINE FORMATION IN PHYSICAL EXERCISE

IN the decades of good hopes for histamine as an essential physiological vasodilator agent, initiated by the pioneer work of Thomas Lewis (1927), the standard experimental procedure was to submit skeletal muscles to anoxia or to a spell of contraction. Release of histamine, which was assumed to occur, could at that time have been searched for by two different approaches: by the demonstration of a decrease in the histamine content of the muscle, or by detecting an increase in the venous blood collected from the muscle. Encouraged, perhaps, by the successful demonstration of the appearance of acetylcholine in the blood draining the structures in which it is released, this latter approach only has been applied in the search for histamine release in muscles.

Emmelin (1945) has recorded that on ultrafiltrating plasma from the guinea-pig, rabbit and rat, a solution was obtained the histamine content of which could be directly assayed on isolated guinea-pig ileum. Duner, Pernow and Tribukait (1958) exposed guinea-pigs to hypoxia corresponding to an altitude of 6000 m and recorded a significant rise in blood histamine with a maximum at 48 hr. In healthy persons performing work on a bicycle ergometer Duner and Pernow (1958) found an increase in histamine in serum, from 1·6 μg/100 ml. whole blood to 2·9 μg during work; in their experiments the number of leukocytes increased from 4400/mm³ blood to 6600 during work. Garden (1966) determined plasma histamine concentration in Marine Corps infantrymen running on a treadmill. An increase in plasma histamine was recorded when the men were running in a hot environment (98° Fahrenheit), but no increase occurred on exercise at room temperature.

In our experiments we explored, in the first place, changes in histamine in the muscle itself, rather than in the blood draining the muscle. In exercise and hypoxia we studied changes in the rate of histamine formation in the muscle tissue proper in three species, cats, rats and mice. Exercise consisted in running in a treadmill for various periods of time. Exposure to reduced O_2 tension was obtained by keeping the animals in a high-altitude chamber at pressures corresponding to 5000 or 6000 m altitude (85–75 mm O_2 pressure). These experiments were carried out by Graham, Kahlson and Rosengren (1964). Embarrassingly, and inexplicably, as in some other instances in histamine research, we encountered species differences.

Mice

Exercise was done in the treadmill for 1 or 3 hr, and lung, skin, hind limb skeletal muscle and diaphragm were excised at 1, 3, 4 and 6 hr after the onset of exercise and their HFC was determined. The results obtained are represented in Fig. 46. Exercise evoked an elevation of HFC with a maximum at about 4 hr. In the lung there was a delay in the onset of elevation and a steep regression. In the three other tissues the changes were more gradual. The elevation was to about 5 times the normal value in the hind limb muscles and about 3 times the normal in the diaphragm, lung and skin.

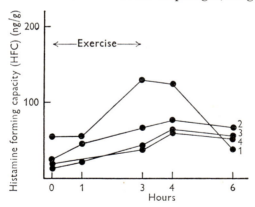

FIG. 46. Histamine forming capacity (HFC) expressed as ng/g tissue in mice exercised continuously for 1 or 3 hr. The tissues, lung (1), skin (2), hind limb skeletal muscle (3) and diaphragm (4), were excised at 1, 3, 4 and 6 hr after onset of exercise. Each point represents the mean of three experiments using pooled tissues from two mice each time. Mean control values of tissues from mice not subjected to exercise are shown at the time 0 (Graham *et al.*, 1964, *J. Physiol.* **172**, 174).

Mice were exposed to oxygen want in a high-altitude chamber and tissues were removed for examination at various time intervals. The values for tissue HFC are summarized in Table 43. On exposure to hypoxia for 6 hr an appreciable elevation of HFC occurred in the skin and muscle which appeared to be returning to normal after exposure for 24 hr. In the lung no significant change of HFC was observed. After re-establishing the normal pressure there was a tendency for the HFC of skin and muscle to remain raised for the period of observation.

Table 43. HFC (ng/g) of normal mouse tissues, after exposure to anoxia for varying periods and at various times after the end of exposure. Each value refers to pooled tissue from two mice

		Lung	Skin	Skeletal muscle
Controls		40	18	15
		62	46	11
		73	15	11
		58	22	12
		74	—	12
Time of exposure to	2 hr	98	27	15
reduced pressure		—	37	9
equivalent to	6 hr	70	81	24
6000 m altitude		84	77	26
	10 hr	44	96	40
		78	58	27
	24 hr	54	42	7
		113	62	9
Time after cessation	2 hr	74	57	14
of exposure to re-		71	57	8
duced pressure	6 hr	58	52	22
equivalent to		88	65	19
6000 m altitude	24 hr	146	37	19
for 24 hr		66	11	7

(Graham *et al.*, 1964, *J. Physiol.* **172**, 174.)

Demedullated mice. Since it is known that during exercise catechol amines are released, we determined the changes in tissue HFC in demedullated mice subjected to exercise. The actual experiments were done 2 months after the operation and the absence of chromaffin tissue was confirmed at autopsy. The demedullated mice ran continuously for 3 hr in the treadmill, and the tissues were examined 4 hr after the onset of exercise because this was the time of peak elevation of HFC in normal mice after similar exercise. The results, which are summarized in Table 44, indicate that in demedullated mice exercise fails to produce significant changes in the HFC of the tissues examined.

Table 44. HFC (ng/g) in tissues of normal and 'demedullated' mice exercised continuously for 3 hr; determinations made 4 hr after onset of exercise. Values from mice not subjected to exercise are included as controls. Each value refers to pooled tissue from two mice

	Lung	Skin	Skeletal muscle	Diaphragm
Controls	40	18	15	19
(unexercised)	62	46	11	19
	73	15	11	—
	58	22	12	19
	74	—	12	—
Intact medulla	149	69	75	79
	100	74	63	42
	130	82	59	51
Demedullated	72	46	24	22
	43	29	15	13

(Graham *et al.*, 1964, *J. Physiol.*, **172**, 174.)

Adrenaline and allied compounds. In order to see to what extent the changes in tissue HFC during exercise and hypoxia were due to catechols discharged into the blood stream, the effects of adrenaline and its congeners were investigated. In mice, the smallest dose of adrenaline given, 2 μg/mouse subcutaneously, evoked a two-fold elevation of HFC in lung, skin, hind limb skeletal muscle and diaphragm when it was measured 3 hr after the injection. Increasing the dose of adrenaline to 5, 10 and 20 μg/mouse, resulted in graded elevations of tissue HFC which at the highest dose levels followed a time course illustrated in Fig. 47. The similarity to the changes described in the corresponding tissues in exercise (Fig. 46) is apparent. Our results with adrenaline injections in mice agree with observations reported by Schayer (1960).

A feed-back relation between HFC and histamine content is discussed on other pages of this book. It appeared conceivable that the elevation of HFC produced by adrenaline might be a sequel to the release and the consequent lowering of tissue histamine

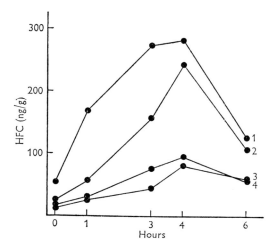

FIG. 47. HFC of lung (1), skin (2), hind limb skeletal muscle
(3) and diaphragm (4) of mice at various times after subcutaneous injection of 20 µg adrenaline/mouse. Each point represents
the mean of three experiments using pooled tissues from two
mice for each determination. Mean control values are shown at
time 0 (Graham *et al.*, 1964, *J. Physiol.* **172**, 174).

content by this agent. We injected adrenaline, 20 µg/mouse subcutaneously and determined the histamine content of lung, skin
and hind limb muscle. From Table 45 it will be seen that adrenaline injections did not produce a lowering of tissue histamine
content detectable at this interval of time by the method used. It
has been claimed by some investigators that adrenaline is a
histamine releaser. The experiment summarized in Table 45 does
not support this claim.

Wherever feasible, changes in tissue HFC observed *in vitro*
should be corroborated by corresponding changes *in vivo*. With
this in view, the whole-body HFC, as reflected in the urinary
histamine, was followed in mice injected with adrenaline. The
urinary excretion of histamine was found increased (Fig. 48) as
should be expected from the elevation of tissue HFC illustrated in
Fig. 47.

Noradrenaline, 20 µg subcutaneously, produced changes in
tissue HFC similar to those obtained with adrenaline, but of a
lower magnitude.

Table 45. Histamine content (µg/g) in tissues of normal mice and 20 min after injection of 20 µg adrenaline/mouse. Each value refers to pooled tissue from three mice

	Lung	Skin	Skeletal muscle
Mice of strain A			
Controls	0·6	78	7·3
Adrenaline	0·7	87	8·8
Mice of strain B			
Controls	0·4	37	4·9
	0·5	35	3·3
Adrenaline	0·5	31	2·3
	0·5	36	3·5

(Graham *et al.*, 1964, *J. Physiol.* **172**, 174.)

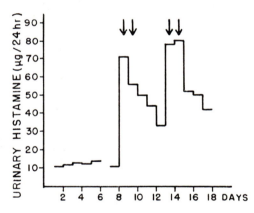

FIG. 48. Urinary histamine (µg/24 hr) in a female mouse injected with 20 µg adrenaline subcutaneously 4 times daily on the days indicated by arrows (Graham *et al.*, 1964, *J. Physiol.* **172**, 174).

Nethalide (promethalol) is credited with the ability to abolish all vasodilator manifestations of injected adrenaline by blocking the β-receptors for adrenaline (Black and Stephenson, 1962; for further references see Whelan and de la Lande, 1963). It appeared interesting to see whether nethalide would abolish the elevation of tissue HFC which is produced by adrenaline. The effect of nethalide alone was investigated in mice by injecting 5 mg/kg slowly intravenously and determining the HFC of excised tissues. The results are given in Table 46, which also includes observations on changes in HFC produced by adrenaline injected 10 min after the administration of the β-blocking agent. Nethalide by itself evoked a substantial elevation of the HFC in the tissues investigated and did not influence the response to adrenaline.

Because of the results with noradrenaline and nethalide, other compounds of similar structure occurring naturally in the body were investigated. Mice were given intraperitoneal injections of DOPA or dopamine, 50 mg/kg, and tissue HFC was determined as in the experiments with adrenaline and noradrenaline. The results are represented in Table 46 which shows that both these compounds induce elevation of tissue HFC in mice.

Table 46. HFC (ng/g) of tissues from normal mice and at 3 hr after injection of adrenaline (20 μg/mouse), nethalide (5 mg/kg), nethalide and adrenaline, DOPA (50 mg/kg) or dopamine (50 mg/kg). Each value refers to pooled tissues from two mice

	Lung	Skin	Skeletal muscle
Controls	40	18	15
	62	46	11
	73	15	11
	58	22	12
	74	—	12
Adrenaline	329	—	40
	194	—	27
	270	242	78
	293	350	85
Nethalide	153	96	42
	111	80	30
Nethalide plus adrenaline	292	162	63
	216	146	130
DOPA	218	113	40
	132	127	51
Dopamine	141	209	74
	115	120	64

(Graham *et al.*, 1964, *J. Physiol.* **172**, 174.)

Exercise, injection of adrenaline or its congeners in mice, all evoked conspicuous elevations of HFC in the four tissues examined, but oxygen want produced an elevation in muscle and skin only. It is noteworthy that the HFC of skeletal muscle at rest was low and that the increase in the rate of histamine formation in exercise was proportionally larger in the muscle than in the other tissues investigated. The HFC remained elevated for at least 3 hr after the mouse has come to rest (Fig. 46). During this period anabolic processes of restoration take place. In considering the possibility of elevated HFC as concerned in the vascular effects of adrenaline, it should be emphasized that this agent accelerates synthesis of histamine in both skeletal muscle and skin where the effect of adrenaline is respectively vasodilator and constrictor. The fact that adrenaline as well as noradrenaline cause elevation of HFC in each of the mouse tissues which we have investigated indicates that the newly synthesized histamine may also be associated with other processes than the control of blood flow, presumably metabolic processes the nature of which remains yet to be disclosed. The experiments with adrenaline in mice were repeated by Pearlman and Waton (1966) who confirmed our results.

Rats

Exercise and hypoxia. Rats were subjected to intermittent exercise for a total of 16 hr, whereafter the lung, skin and skeletal muscle were removed and their HFC determined. The results are summarized in Table 47 which shows a striking fall of HFC in the lung, whereas in the skin and muscle no obvious change occurred.

For urine collection rats were forced to walk in a treadmill fitted into a standard metabolism cage. They were adequately fed by stomach tube and given subcutaneous injections of saline to maintain a steady urine volume. A typical experiment is illustrated in Fig. 49, which shows that exercise lowers the urinary histamine in proportion to the duration of exercise.

In rats exposed to oxygen want (reduced pressure equivalent to 5000–6000 m altitude), there was a tendency for the urinary histamine to fall. The tissue HFC, on exposure to oxygen want, was altered as follows: in hind limb skeletal muscle and abdominal skin the changes were inconsistent and insignificant, and in the lung a fall occurred to about 30% of normal within 2 hr of exposure and to about 15% at 6 hr. This fall in the HFC of the lung could account for the small reduction in urinary histamine described above.

Table 47. Histamine forming capacity (HFC), expressed as ng/g, of normal rat tissues, after periodic exercise (5 min walk, 4 min rest for 16 hr), adrenaline, or cortisone. Tissues examined at the end of exercise, at 3 hr after injection of adrenaline, and at 4 hr after injection of cortisone

	Lung	Skin	Skeletal muscle
Controls	272	34	7
	366	70	12
	218	15	5
	602	20	8
	556	—	9
	542	—	12
	636	19	8
	420	19	11
Exercise	85	16	13
	90	11	6
	94	14	5
Adrenaline (μg/rat)			
5	128	44	7
5	129	35	13
20	126	26	6
40	154	40	17
40	92	6	5
40	172	27	6
100	81	48	11
100	92	58	12
Cortisone (mg/rat)			
1	79	15	5
1	78	15	5
1	78	14	6

(Graham *et al.*, 1964, *J. Physiol.* **172**, 174.)

Adrenaline. In rats we investigated various dose levels of adrenaline, 5–100 μg/rat subcutaneously, and the tissues were examined 3 hr after injection. The results are included in Table 47 from which it will be seen that the doses of adrenaline employed produced a considerable fall in the HFC of the lung and no obvious changes in skin and muscle.

Cortisone. Schayer and his colleagues have shown that administration of glucocorticoids in rats causes changes in histidine decarboxylase activity of various tissues (Schayer, 1956*b*). In rats

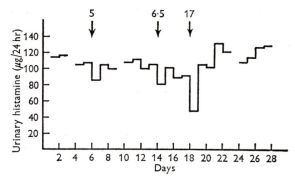

Fig. 49. Urinary histamine (μg/24 hr) in a female rat exercised at the days marked by arrows for the total number of hours indicated above the arrows. The exercise consisted of periods of 6 min walking and 3 min rest. The rat received aminoguanidine 20 mg/kg per day subcutaneously (Graham *et al.*, 1964, *J. Physiol.* **172**, 174).

this enzyme was reported to be significantly inhibited in the lung and liver, but in the mouse lung no inhibition was detected by Schayer and Ganley (1961). It seemed of interest to investigate the effect of cortisone in the strains of rat which we studied because these effects might provide an explanation for the fact that the changes in tissue HFC during exercise and anoxia were different from those in mice. On injecting 1 mg cortisone intramuscularly the HFC at 4 hr after the injection was substantially reduced in the lung, whereas in skin and muscle no obvious changes were seen (Table 47).

Cats

Presence of histidine decarboxylase. As recently as in 1963 it was maintained by some investigators that 'in cats the evidence for the presence of tissue histidine decarboxylase is anything but convincing' (Waton, 1963). This view was based on determinations with the non-isotopic method in which benzene is added. As we have already discussed, benzene is alleged to inhibit histidine decarboxylase in a method which is inherently already lacking in sensitivity. With this method Kameswaran and West (1962) reported the absence of histidine decarboxylase in all feline tissues examined, and Waton (1962) found no significant formation of histamine in any of five feline tissues, from 18 days before birth to 54 hr after birth. Waton concluded that the foetal tissues and the tissues of very young kittens behaved like the tissues of the adult cat which he had examined in 1956. However, employing Schayer's isotopic methods, Waton (1964) found ^{14}C-histamine in all tissues examined after an intravenous infusion or injection of ^{14}C-histidine.

The HFC of feline tissues is low, indeed, and can be safely determined isotopically only if the background values are kept very low by the procedures described in Chapter 5. Representative HFC values are given in Table 48.

Table 48. HFC (ng/g) of tissues of normal kittens and at 3 hr after injection of adrenaline

	Lung	Skin	Skeletal muscle	Kidney
Controls	10	17	4	2
	20	10	3	5
	20	11	5	4
Adrenaline 80 µg/100 g	17	10	3	3
	9	21	1	2
Adrenaline 120 µg/100 g	8	5	5	4
	4	5	3	5

(Graham *et al.*, 1964, *J. Physiol.* **172**, 174.)

Adrenaline. We examined the effect of adrenaline in kittens, untreated litter mates being used as controls. The results in seven kittens are presented in Table 48. Adrenaline, 80 µg/100 g, administered to kittens subcutaneously caused no, or only inconsistent and insignificant, alterations in the HFC of four tissues investigated 3 hr after the injection. A higher dose, 120 µg/100 g subcutaneously, evoked a fall in HFC of the lung and skin. The fall in the lung is noteworthy considering Schayer's (1962*b*) finding of increased histidine decarboxylase activity in the lung of cats after injection of endotoxin.

Comments: The discrepancies in the results obtained in physical exercise and hypoxia in mice, rats and cats appear bewildering. First, there are the species differences, the tissue HFC changing in opposite directions in mice and rats. Next, humoral agents known to be released in exercise and anoxia, notably catechol amines, produce changes in tissue HFC similar to those occurring in exercise. The changes seen in exercise in mice appear physiologically meaningful. Increased histamine formation has been proposed to be related to metabolic processes and may contribute to the vasodilatation and to the activation of the pituitary–adrenal system which take place in exercise. It remains to be discovered whether or not in exercise in man changes in tissue histamine similar to those in mice exist. The profound changes in histamine metabolism which have been recorded in exercise, the meaning of which are at present not clear, are likely to provide a field for further extensive studies.

12

FORMATION AND METABOLISM OF HISTAMINE IN THE BRAIN

IN the course of studies on histamine and its relation to traumatic shock Kwiatkowski (1943) found relatively large amounts of histamine in extracts of mixed somatic nerves which to him, and subsequently to others, suggested the existence of histaminergic nerves. He also assayed extracts of the cortex, mid-brain and cerebellum in various laboratory animals, but the presence of histamine could be detected in the cerebellum only. There was little reason for further exploration of the histamine content of the brain until it was discovered that the hypothalamic regions convey stimulatory influence to the anterior pituitary gland via the hypophyseal–portal vessels. It became apparent that the hypothalamus controls the functions of the anterior pituitary gland by releasing regulatory substances into the portal blood flowing from the hypothalamus into the pars distalis of the gland (for references, see Harris, 1955). At the time of the discovery of this portal blood system by Wislocki, more than thirty years ago, naturally nothing was known about the possible chemical nature of the stimulant released by the hypothalamic regions. Any normal constituent of the brain known to be stimulatory had therefore to be considered. Since we know how to extract and assay histamine, and as the problem eventually had to be dealt with, the occurrence and distribution of histamine in cerebral regions related to the hypophysis was investigated by Harris, Jacobsohn & Kahlson (1952). The median eminence of cats, dogs and pigs was found to be conspicuously rich in histamine and in several instances its histamine concentration was of the same order as in the gastric mucosa. The median eminence is the region into which stimulants for the pars anterior are believed to gain entrance. The pars anterior proper was also found to be rich in histamine. The hypothalamus contained considerable concentrations, approximately the same as in sympathetic ganglia. The concentration in the thalamus, nucleus caudatus and cerebellum was very low. It has been emphasized by Adam (1961) that the pattern of distribution of histamine as outlined by Harris *et al.* closely resembles the distribution described by Vogt (1954) for noradrenaline and by Amin, Crawford and Gaddum (1954) for 5-hydroxytryptamine.

In the study by Harris *et al.* considerable concentrations of histamine were found in both the anterior and posterior lobes of the hypophysis. Similar results were obtained by Adam and Hye (1966) who in addition showed that the histamine found in the hypophysis derives partly from mast cells which occur chiefly in the posterior lobe and the pars tuberalis of the adenohypophysis. These authors find it probable that histamine extractable from these tissues derives mainly from mast cells. The hypothalamus is generally believed to be devoid of mast cells.

Having discovered that certain cerebral regions were rich in histamine, complying with our general belief, it remained to demonstrate that the cerebral regions which contain histamine also produce it. This arduous task was undertaken by our colleague White. Histamine formation by brain tissue *in vitro* had been demonstrated in cattle (Holtz & Westermann, 1956) and in the rat by Schayer (1956*b*).

White (1960) accomplished the feat of perfusing the cat's brain and measuring the rate of formation of ^{14}C-histamine from ^{14}C-histidine. He further showed that in the brain *in vivo* histamine was converted to methylhistamine and methylimidazole acetic acid and that the formation of this latter compound was inhibited by iproniazid.

These experiments *in vivo* were preceded by investigations *in vitro*. White (1959) determined the rate of ^{14}C-histamine formation from ^{14}C-histidine in various regions of the brain from cats, dogs and pigs. He found the highest rate of histamine formation in hypothalamic tissue. In the median eminence the HFC was also relatively high, 214 ng/g. In pooled tissue from five pigs the histamine formation was virtually nil in the anterior pituitary and 12 ng/g in the posterior pituitary. In the medulla oblongata tissue from three cats it was 3–5 ng/g and in one cat pituitary, nil. Values obtained in various brain regions of three species are given in Table 49.

G

Table 49. Histamine formed in 3 hr by brain tissue in N_2 from 40 μg ^{14}C-histidine (expressed as ng/g tissue)

Species	Cortex (2·00 g)	Cerebellum (1·40–2·00 g)	Hypothalamus (0·25–0·75 g)	Area postrema (0·25 g)
Cat	14	≪2	191	—
Cat	11	≪2	141*	—
Cat	11	≪2	134*	—
Cat	6	2	333	—
Pig†	17	0	—	—
Dog	10	—	355	50
Dog	14	—	1116	23

* Pooled tissue from 2 cats. † pooled tissue from 5 pigs.
(White, 1959, *J. Physiol.* **149**, 34.)

Histamine binding. On incubating tissue samples with ^{14}C-histidine and centrifuging, the greater portion of ^{14}C-histamine formed in the hypothalamus was found in the supernatant fluid after centrifuging for 5 min only, indicating lack of binding sites and that the histamine formed was non-mast cell in nature. The loosely bound histamine leaves the site of its formation and, in so far it is not locally catabolized, can be carried to distant sites to serve specific functions there, i.e. stimulate the anterior pituitary gland as will be discussed in Chapter 13.

Histamine catabolism in brain tissue

The figures for histamine formation given in Table 49 should be regarded as minimum values, as it is possible that formation and catabolism of histamine proceed simultaneously. The catabolism was investigated by White (1959, 1960, 1964) who incubated brain tissue with added ^{14}C-histamine under conditions identical with those of his histamine-formation experiments whereafter the metabolites of histamine were determined by isotope dilution methods. As is shown in Table 50, brain tissue *in vitro* transforms histamine into ring-N-methyl derivatives, whereas the oxidation of histamine to imidazoleacetic acid is insignificant. This latter fact indicates that there is no histaminase (diamine oxidase) in the brain, a fact which has been confirmed by Burkard, Gey and Pletscher (1963).

In the brain *in vivo* the pattern of catabolism was the same as *in vitro*. This was shown by employing a technique similar to that

Table 50. Histamine metabolism in cat brain tissue; 3 hr at 37°C; 2μg ^{14}C-histamine added to each sample. The figures denote the percentage of added histamine recovered as histamine and metabolites

Gas phase	Tissue	^{14}C-histamine	^{14}C-methyl-histamine	^{14}C-methyl-imidazole-acetic acid	^{14}C-imidazole acetic acid	Total recovery
N$_2$	C*	32	70	≤1	0	103
N$_2$	C	31	71	≤2	0	105
O$_2$	C	23	66	13	≤1	102
O$_2$	C	14	60	11	0	85
N$_2$	H	74	20	≤1	0	95
N$_2$	H	73	19	≤1	0	94
O$_2$	H	82	20	5	0	106
O$_2$	H	64	20	4	≤1	90

* C, cortex 2·00 g; H, hypothalamus 0·25 g.
(White, 1959, *J. Physiol.* **149**, 34.)

devised by Bhattacharya and Feldberg (1958) in which the lateral ventricle was cannulated in anaesthetized cats and the perfusion carried out from the lateral ventricle through the cavities of the brain, to the aqueduct, or to the cisterna magna. When perfusing radioactive histamine this way, a small fraction of it entered the brain tissue. Practically all of the histamine that entered the brain tissue was methylated and recovered as methylhistamine or methylimidazoleacetic acid. There was no measurable formation of imidazoleacetic acid. These results are in agreement with observations by Brown, Tomchick and Axelrod (1959), who found that brain tissue is particularly rich in the enzyme imidazole-N-methyl transferase. In White's experiments pretreatment of the cats with iproniazid completely inhibited the formation of methylimidazoleacetic acid, but not the formation of methylhistamine. Iproniazid is a monoamine oxidase inhibitor which blocks the oxidation of methylhistamine to methylimidazoleacetic acid (Rothschild & Schayer, 1958b). In White's experiments (1966) treatment of cats with this compound led to the accumulation of methylhistamine in the brain (Fig. 50, Table 51). Reserpine had variable effects on hypothalamic methylhistamine, and the histamine concentration of this region also varied after reserpine, but it was consistently lower than in the controls. In the other regions examined no significant deviations from the control values were observed (Table 51). Adam and Hye (1964) have reported that in cats reserpine reduced the concentration of histamine in the hypothalamus.

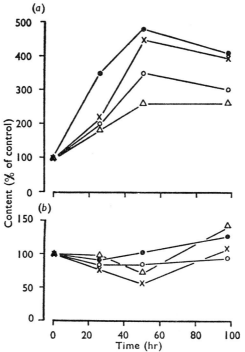

FIG. 50. Methylhistamine (*a*) and histamine (*b*) in cat brain 26, 50, and 98 hr after the beginning of treatment with iproniazid (25 mg/kg, daily). Ordinate: concentrations, as percentages of controls (Table 51). Each point represents the mean of two experiments. ●, Hypothalamus; ○, caudate nucleus; ×, thalamus; and △, mesencephalon (White, 1966, *Br. J. Pharmac.* **26**, 494).

Further work on methylhistamine in the brain. White (1966) developed a non-isotopic method for the simultaneous measurement of the content of histamine and methylhistamine in the brain and other tissues. It is based on the formation of the dinitriphenyl derivatives of the amine, a reaction described by others who are cited by White. Employing this procedure, White investigated the cat brain and other tissues, and in particular the regions of the brain in which Harris *et al.* (1952), and Adam (1961) and Adam and Hye (1966), in more detailed studies, had found the presence of high concentrations of histamine. Among the tissues

Table 51. Concentrations (µg/g) of histamine and methylhistamine in different regions of the brain stem in the cat. Control values are means and standard errors

Histamine and methylhistamine (µg/g) for expt. no.

Tissue	1–8 Controls	9, 10 Iproniazid 3×25 mg/kg	Reserpine 11, 12 0·4 mg/kg	Reserpine 13, 14 1·0 mg/kg	Chlorpromazine 15, 16, 17 4×20 mg/kg	Chlorpromazine 18 8×20 mg/kg	Chlorpromazine 19 2×50 mg/kg
Histamine							
Hypothalamus	0·93 ±0·11	0·8, 1·1	0·0, 0·7	0·3, 0·5	1·1, 1·2, 1·3	1·3	1·1
Caudate nucleus	0·34 ±0·05	0·2, 0·3	0·0, 0·4	0·3, 0·2	0·2, 0·3, 0·3	0·5	0·4
Thalamus	0·34 ±0·11	0·2, 0·2	0·3, 0·4	0·7, 0·5	0·2, 0·3, 0·3	0·4	0·4
Mesencephalon	0·21 ±0·07	0·2, 0·1	0·1, 0·3	0·3, 0·1	0·1, —, 0·2	0·2	0·4
Methylhistamine							
Hypothalamus	0·56 ±0·12	2·8, 2·6	0·3, 1·7	0·1, 0·6	0·7, 0·2, 0·5	0·8	0·3
Caudate nucleus	0·40 ±0·05	1·7, 1·1	0·3, 0·6	0·2, 0·2	0·5, 0·2, 0·3	0·3	0·5
Thalamus	0·30 ±0·05	1·7, 1·0	0·1, 0·3	0·1, 0·1	0·1, 0·0, 0·2	0·2	0·3
Mesencephalon	0·27 ±0·09	0·7, 0·7,	0·1, 0·5	0·1, 0·2	0·1, —, 0·2	0·2	0·1

(White, 1966, *Br. J. Pharmac.* **26**, 494.)

listed in Table 52, the hypophysis and the gastric mucosa had the highest content of methyl histamine.

Table 52. Concentrations of histamine and methylhistamine in some cat tissues. Each value represents the mean of three experiments, except for the hypophysis which represents one determination on pooled tissue from six cats

Tissue	Concentration of	
	Histamine (µg/g)	Methyl-histamine (µg/g)
Cerebral cortex	0·1	0·3
Cerebellar cortex	0·1	0·1
Hypophysis	1·8	2·2
Lung	12·8	0·7
Kidney	0·1	0·2
Liver	1·2	0·2
Spleen	2·0	0·5
Gastric mucosa	27·7	1·1

(White, 1966, *Br. J. Pharmac.* **26**, 494.)

Among brain regions the hypothalamus has from the onset been our main interest. In the experiments of White (1966) the hypothalamus was found to contain less methylhistamine than histamine, whereas in other cerebral tissues examined the concentrations of the two amines were approximately equal (Table 51). It has already been explained that in the brain the rate of increase of methylhistamine can be taken to reflect the rate of its formation, and also the turnover rate of histamine. Here, again, it is shown that there is a higher turnover rate for histamine in the hypothalamus than in the other regions examined. In passing, results obtained by Jonson and White (1964) should be cited indicating that in the brain exogenous histamine is more rapidly methylated than endogenous histamine.

Inhibition of methylation of brain histamine. It has been explained that methylation seems to be the major pathway for histamine catabolism in the brain. The first successful attempt to inhibit methylation of histamine in the living brain was made by White (1961). At that time Brown, Tomchick and Axelrod (1959) had shown that a purified preparation of the histamine-methylating enzyme was inhibited *in vitro* by chlorpromazine, although rather weakly, and Lindahl (1960) had reported inhibition of this enzyme *in vitro* by cupric ions. White perfused the brain of cats from the lateral ventricle to the aqueduct. The effects of copper and of intramuscular chlorpromazine in aqueductal perfusions are summarized in Fig. 51. The blocks represent the total methylation, that is, the sum of methylhistamine and methylimidazoleacetic acid. The first five open blocks represent untreated cats, and the others represent cats which had received $CuCl_2$, or chlorpromazine. White remarks that it is not possible to claim from these results that there is a specific effect of chlorpromazine on the methylating enzyme.

In the later *in vitro* study on histamine and methylhistamine in cat brain White (1966) found that chlorpromazine had little or no

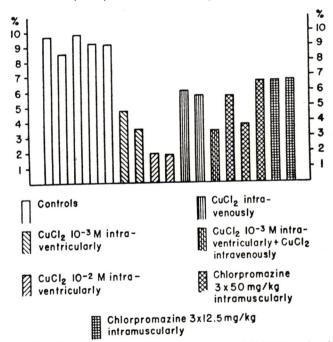

FIG. 51. Diagram to show the methylation of ^{14}C-histamine in cat brain during perfusions from the lateral ventricle to the aqueduct. Percentage of infused ^{14}C-histamine recovered as methylhistamine + methylimidazoleacetic acid. Each column represents one experiment (White, 1961, M.D. Thesis; 1964, *Fedn. Proc.* **23**, 1103).

effect on the amines studied, with the exception of hypothalamic histamine, which was significantly increased. This latter finding agrees with results of Adam and Hye (1964) on the cat hypothalamus, whereas in the whole brain of the rat increased histamine concentration has been reported by Green and Erickson (1964). This latter finding is suggestive of species differences.

Possible roles for histamine formation in the brain. The discovery, about twenty years ago, of a differential distribution of histamine in the brain and its accumulation in functionally strategic regions, followed by the demonstration that histamine is formed preferentially in these regions, has opened up a new field of research. Fortunately, this field has attracted a number of able investigators. This is not the place to cite or discuss their findings. References to their work will be found in the Rocha e Silva Handbook (1966) and in the Symposium: *Histamine and the Nervous System*, edited by Clark and Ungar (1964).

In the opinion of the present authors, histamine formation in the brain is associated with more than one function, the most obvious ones being a role in the machinery of transmission and in anabolic processes. However, there is no evidence for this. We do not believe that injections of histamine, or exposing the brain, or parts of it, to extracellular histamine, is likely to provide useful information. A straightforward approach to the problem of the function of brain histamine would be to inhibit histamine formation in descrete regions of the brain. Obviously, this cannot at present be done and it is likely to be impracticable for a long time.

Brain histamine should on no grounds be equalled to nascent histamine for which one precise role has been proposed, the anabolic one. Nevertheless, there are resemblances between the two kinds: the brain histamine under discussion is non-mast cell in nature and it disappears rapidly from the site of its formation. Perhaps a clue in the search of the functions of histamine in the brain lies in the fact that it is preferentially formed in regions which exercise distinct physiological functions.

In conclusion, we feel alliance with Erspamer (1961) in his remark: 'Histamine in the CNS has been ignored for a long time by several investigators as a second-class amine. But this amine, however annoying the fact may be, has the same citizenship rights in the CNS as catecholamines and 5-HT, whose function in the CNS is approximately as obscure as that of histamine.'

13

HISTAMINE AS A STIMULANT TO THE ANTERIOR PITUITARY GLAND

HISTAMINE is differentially concentrated, and produced, in the hypothalamus and the median eminence (Harris, *et al.*, 1952; White, 1959). The regulation of secretion of pituitary hormones is dependent on the integrity of its hypophysial portal-blood supply. These vessels arise as capillaries in the median eminence, whence blood is carried to the anterior lobe of the pituitary gland (for references see the Monograph by Harris, 1955). It would appear that these vessels carry chemical agents which convey instruction to the anterior pituitary gland, which in turn, responds by secreting the particular trophic hormone that is appropriate to the nature of the stimulating agent. The humoral stimulants are believed to be peptides. Bullock median-eminence extract produced the ovulation response in rabbits after intrapituitary infusion of the extract. The releasing factor of this extract appeared to be a basic peptide which produced ovulation at a dose of 6 μg of peptide/animal (Fawcett, Reed, Charlton & Harris, 1968). These investigators were not in the fortunate position of Gregory and his colleagues, who in purifying and isolating gastrin could avail themselves of large bulks of tissue for extraction.

Figure 52 from the report by Fawcett *et al.* (1968) shows that the capillaries of the portal blood flow to the anterior pituitary gland arise in regions which are rich in histamine and endowed with the capacity to form it. Histaminase (diamine oxidase) appears to be absent in these regions. Histamine formed therein is loosely bound and it diffuses freely. It has been reported that the pituitary gland is unable to form histamine, but that the gland contains it. This situation, then, stands as an exception from the generalization that histamine containing tissues produce their own histamine. Histamine contained in the pituitary gland has been shown to be taken

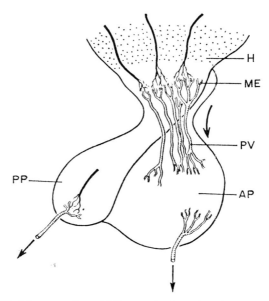

Fig. 52. Diagram of a mid-line sagittal section through the base of the hypothalmus and pituitary gland to show the structures concerned in the regulation of anterior-pituitary activity. Nerve fibres from the hypothalamus (stippled; H enter the median eminence (ME) of the tuber cinereum to end on the primary plexus of the hypophysial portal vessels (PV). It is postulated that releasing factors pass from these nerve terminals into the portal vessels and are carried to the anterior pituitary gland (AP) to regulate its secretory activity. This is probably a similar mechanism to that obtaining in the posterior pituitary gland (PP), where peptides are discharged from nerve endings into the general circulation (Fawcett *et al.*, 1968, *Biochem. J.* **106**, 229).

up from the blood (Adam, Hye & Waton, 1964). Further, it has been shown that histamine carried in the blood is capable of stimulating the gland to secrete ACTH.

A study of blood-borne histamine as a stimulant for the anterior pituitary gland will now be discussed (Fuche & Kahlson, 1957). These authors worked on unanaesthetized rabbits, normal, hypophysectomized or deprived of the adrenal medulla. The degree of incited release of ACTH was judged by the lymphopenic response which in the unanaesthetized rabbit has been proved of value as an indicator of anterior pituitary activity (Colfer, de Groot & Harris, 1950; de Groot & Harris, 1950). ACTH (Armour Products) and an ACTH protein donated by Dr. C. H. Li were employed. Either preparation in a dose of 0·05 mg/kg intravenously gave definite lymphopenic responses.

The amounts of histamine infused in an ear vein were so chosen as to give responses which were approximately equal in magnitude to those obtained with ACTH. The smallest amount of histamine to give a significant lymphopenic response in most cases was 25 μg/kg (in terms of the base). In its effect this dose was roughly comparable to 0·025 mg ACTH/kg. It was therefore decided to use histamine, 50 μg/kg as a standard dose, and 100 μg/kg for additional information. Such doses are likely to augment the discharge of adrenaline from the adrenal medulla (Burn & Dale, 1926). Adrenaline, in turn, induces a lymphopenic response. For this reason the adrenal medulla was removed in some experiments. Destruction of the adrenal medulla neither abolished nor significantly altered the lymphopenic response to injected histamine. It thus appears that the lymphopenic response to histamine was not mediated by liberated adrenaline. Hypophysectomy was carried out in rabbits serving as controls. From the second week after the operation the intact and hypophysectomized groups were fully comparable as regards pre-injection blood lymphocyte level.

The main results obtained in this study are the following: (1) Intravenous infusion of histamine (in most cases 50 μg/kg body weight) caused a lymphopenic response of the same type and time-course as that following an injection of ACTH. (2) After hypophysectomy histamine failed to give distinct lymphopenic responses. (3) Adrenaline, released by the administration of histamine, was not essential in the lymphopenic response to histamine. (4) The histamine antagonist, mepyramine, did not abolish the lymphopenic response to histamine.

Histamine may reach the anterior pituitary gland by two routes: (1) From the cerebral regions in which the capillaries of the portal blood supply arise. The components and the arrangement of this region are precisely so designed as to carry the histamine formed therein to the pituitary gland. This circumstance appears so obvious that dissenting views on this point have so far not been placed on record. The occurrence, on adequate stimulation, of histamine in the portal blood, might be demonstrated directly, since techniques have been developed to collect blood from the hypophyseal portal vessels and because methods are now available to detect even ng of ^{14}C-histamine formed in the pertinent cerebral regions after injection of ^{14}C-histidine. Assuming that this experiment will be done, the problem would remain unresolved as to whether the concentration of histamine in the portal blood is sufficiently high enough to serve as a stimulant to the gland. (2) Regarding histamine carried by the systemic blood, the stimulatory active concentration has been established in the study described above. Histamine is known to be released and, moreover, to be newly formed in a great variety of conditions. The amounts of the amine which hereby enter the systemic circulation may be large. In the rabbit, for example, in antigen–antibody reactions, about 300 μg/kg of histamine is released (Schachter, 1953). In previous chapters we indicated that great amounts of histamine are newly formed (elevated HFC) in several normal conditions. This is mainly non-mast-cell histamine, loosely bound, which enters the blood stream and is in part excreted in the form of the amine or its metabolites. With sufficiently high concentrations, it is difficult to conceive why histamine carried by the systemic blood should not stimulate the secretion of ACTH. If this assumption could be strictly proved, blood-borne histamine would in this respect be accorded a rank among the many other guardians of homeostasis.

The topic of histamine as a stimulant of the pituitary–adrenal system has attracted the attention of only a few investigators. Fortier (1951) injected rather large doses of histamine acid phosphate (1 mg/100 g) subcutaneously in rats and employed the eosinopenic response as an index of ACTH release. Histamine injection brought about a definite eosinopenia in both the normal state and when the adenohypophysis had been grafted into the anterior chamber of the eye. Fortier suggests a dual regulation of ACTH release, one by blood-borne humoral agents, the other probably by the hypothalamus–hypophyseal pathways. The results by Fuche and Kahlson (1957) have been confirmed by Suzuki, Hirai, Yoshio, Kurouju and Yamashati (1963), by direct determination of 17-hydroxycorticoid in unanaesthetized dogs. They injected histamine intravenously in doses of 0·05–1·0 mg/kg and recorded a significant increase in the secretion of the corticoid.

In sum, we are far from implying that histamine, cerebral or in the systemic circulation, is the sole stimulant causing the pituitary gland to secrete ACTH. There are other(s) in addition. In the control of functions we often encounter expedients which are alternative, or supporting, to the principal one(s). The most suggestive among those is the histamine formed in the region in which the capillaries of the blood flow to the pituitary gland arise. Low molecular, non-peptide, agents should not *a priori* be excluded in this connexion since the molecular structure of ovine hypothalamic hypophysiotropic TSH-releasing factor has been established as 2-pyrrolidone-5-carboxylyl-L-histidyl-L-proline amide (Burgus, Dunn, Desiderio, Ward, Vale & Guillemin, 1970).

14

HISTAMINE FORMATION IN HYPER-SENSITIVITY STATES

EVER since the work of Bartosch, Feldberg and Nagel, and in the same year, 1932, of Dragstedt and Gebauer-Fuelnegg, it has become common knowledge that histamine is released in anaphylaxis. Mast cells are believed to be the main source from which the histamine is released. Although there is considerable evidence for this, Mota (1966) prudently proposed: 'In conclusion, so far as we can judge from evidence which is by no means yet so complete as we would like it to be, the mast cells play a very important role in guinea-pig anaphylaxis, due to the known high sensitivity of this species to histamine, but probably an insignificant role in rat anaphylaxis due to the low sensitivity of this species to histamine and 5-hydroxytryptamine.'

In anaphylaxis histamine release takes place rapidly. The time course of release has been well established *in vitro*. *In vivo* the period during which histamine is released is less certain, owing to lack of exact criteria of release. The few investigations which provide information on the time course of release *in vivo* are based on determinations of increased blood histamine content. Mongar and Whelan (1953) injected D-tubocurarine into the human brachial artery, arrested the blood flow in the arm for 2 min to allow released histamine to accumulate, and recovered the released histamine in the plasma in the venous return from the arm. The whole process of histamine release was accomplished within a few minutes, as shown in Fig. 53.

In experiments by Schachter and Talesnik (1952) on rats injected only once with egg-white, the plasma histamine concentration increased to a maximum within 6 min, and in a few rats an increase was still present after 60 min. Mota (1966) determined the histamine content of plasma in the state of anaphylactic shock in rats

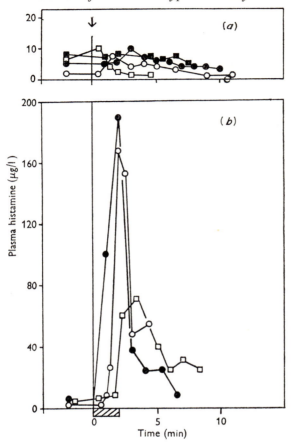

FIG. 53. Effect of intra-arterial D-tubocurarine on plasma hista-mine levels. (*a*) anaesthetized subjects: ○, 5 mg/min for 6 min; □, 10 mg/min for 3½ min. Conscious subjects: ●, 1 mg/min for 8 min; ■, 1 mg/min for 2½ min. The arrow indicates the begin-ning of the infusion in each case. (*b*) anaesthetized subjects, single injections (○, 50 mg; □, 10 mg; ●, 10 mg), followed by circulatory arrest for 2 min (hatched area). The scale in (*b*) is double that of (*a*) (Mongar & Whelan, 1953, *J. Physiol.*, **120**, 146).

and found that the histamine content reached a maximum at about 5 min after the injection of antigen and then declined so that no histamine could be detected 20 min later. Again, *in vivo*, the period of histamine release was relatively brief.

In discussing humoral agents alleged to produce the symptoms of anaphylaxis, whether or not it is histamine, kinins, slow reacting substances, and others, little emphasis is laid on the fact that the symptoms of anaphylaxis in certain species and circumstances may prevail for many hours. Such long-lasting symptoms cannot possibly be sustained by a rapid release of preformed tissue constituents.

Aware of the association between lowering of histamine content (release) and elevation of HFC, and having observed elevated HFC in tissues after damage and during the process of repair (in anaphylaxis cells are damaged), it appeared *a priori* likely that the rate of histamine formation should be increased in anaphylaxis. On these premises Kahlson, Rosengren and Thunberg (1966) investigated tissue HFC and the time-course of changes therein in anaphylaxis in guinea-pigs and rats.

Rats were sensitized with hen egg-white and *Haemophilus pertussis* vaccine as an adjuvant, as introduced by Malkiel and Hargis (1952). The dosage of the second injection of egg-white, 11–13 days later, was so chosen as to produce only mild anaphylactic manifestations. Guinea-pigs were sensitized with hen egg-white according to Mongar and Schild (1957). For re-injection 3 weeks later a sub-lethal dose was employed.

It was noted that in rats *H. pertussis* vaccine by itself caused elevation of HFC in some tissues. A similar occurrence has been recorded in mice by Schayer and Ganley (1959).

The observations in anaphylaxis are for brevity illustrated in Fig. 54. In the anaphylactic reaction the rate at which histamine is newly formed was accelerated in most of the tissues investigated. The time course of these changes was largely the same in rats and guinea-pigs, and there are two rather distinct types of time course. In some tissues peak values of HFC occurred within 3–6 hours, whereas in others peak HFC was reached only after 24 hours. In some tissues the enhanced rate of histamine formation persisted at 48 hours, after which time no observations were made. Numerical values are given in Table 53, which includes HFC in tissues of controls (given adjuvant only), the hour and value for the peak HFC after re-injection, and the statistical significance of the changes. The exceedingly high HFC of rat lung, the delayed peak elevation in rat small intestine and guinea-pig spleen, and the fall in HFC of gastric mucosa, seem particularly noteworthy. The

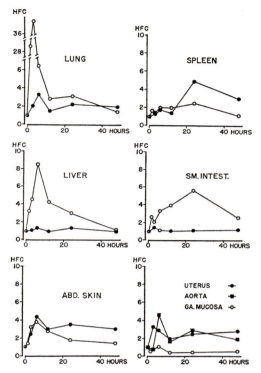

Fig. 54. Histamine-forming capacity (HFC) of various tissues in anaphylaxis in rats (open symbols) and guinea-pigs (filled symbols). Changes produced by reinjection of antigen at zero time are expressed in terms of multiples of control values. Each value represents the mean of results obtained in three to five animals (Kahlson *et al.*, 1966, *Lancet* **i**, 782).

small intestine of rats was filled with blood-stained fluid in anaphylaxis.

In hypersensitivity and anaphylactic reactions of the types described, the tissues are exposed to histamine which originates from two different processes with distinct time courses: an initial brief phase of release of histamine, and the supervening prolonged phase of formation of histamine. The initial liberation is well established, whereas the elicited acceleration of histamine formation is a recent discovery (Kahlson, Rosengren & Thunberg, 1966). The duration of elevated HFC may exceed 48 hours,

Table 53. Histamine-forming capacity (HFC) of rat and guinea-pig tissues from sensitized controls, and at hour of maximal elevation in anaphylaxis (figures denote HFC in ng/g with standard error of the mean, number o f animals in parentheses)

		Anaphylaxis			
		Peak			Signifi- cance of
Tissue	Controls HFC	hr.	Peak HFC		peak
Rat:					
Lung	280 ± 55 (5)	$1\frac{1}{2}$–3	8100 ± 1800	(9)	$p < 0.01$
Liver	83 ± 12 (3)	6	680 ± 38	(4)	$p < 0.001$
Abdominal skin	26 ± 2 (3)	6	72 ± 16	(4)	$p < 0.05$
Spleen	640 ± 61 (3)	24	1580 ± 320	(6)	$p < 0.1$
Small intestine	12 ± 3 (3)	24	66 ± 11	(6)	$p = 0.01$
Gastric mucosa	9600 ± 2600 (5)	12–48	3900 ± 470	(5)	$p < 0.1$
Guinea-pig:					
Lung	34 ± 6 (5)	6	113 ± 17	(4)	$p < 0.01$
Liver	22 ± 5 (5)	6	30 ± 2	(4)	$p < 0.01$
Abdominal skin	3 ± 1 (5)	6	13 ± 11	(4)	$p < 0.01$
Spleen	29 ± 4 (5)	24	141 ± 27	(4)	$p < 0.01$
Small intestine	53 ± 3 (5)	3	78 ± 6	(4)	$p < 0.01$
Uterus	14 ± 2 (5)	3	46 ± 10	(4)	$p < 0.01$
Aorta	18 ± 4 (5)	6	84 ± 11	(4)	$p < 0.001$

(Kahlson *et al.*, 1966, *Lancet*, i, 782.)

and elevations of even greater magnitude are likely to occur on re-injecting larger doses of antigen than those employed in our experiments.

Histamine resulting from elevations of tissue HFC is bound to cause effects in two different ways. Newly formed histamine will in the first place act within the cells and tissues in which it is formed, thereby acquiring a distinct functional status, not to be confused with released 'intrinsic', histamine. The latter is preformed histamine leaving the cell, whereas the newly formed is building up intracellularly at gradually increasing rates. The effects of histamine generated by elevated HFC and acting at the site of its formation are not abolished by histamine antagonists. A portion of the histamine formed at high rates in anaphylaxis will be catabolized before entering the blood stream. The portion entering the blood will exert effects similar to those produced by released or injected histamine. These effects are to some extent counteracted by histamine antagonists.

The guinea-pig deserves special consideration on account of its high sensitivity to histamine and the ease with which hypersensitivity reactions can be produced. In the normal state (no adjuvant) the guinea-pig tissues examined exhibited HFC in order of decreasing magnitude as follows: small intestine, lung, spleen, liver, aorta, uterus, skin. Table 53 answers the question whether tissues with particularly high HFC in the normal state have correspondingly high HFC levels in anaphylaxis, and vice versa. This is not the case. Nor do the tissues in which anaphylactic manifestations are predominant, for example, guinea-pig lung and rat intestine, exhibit strikingly high elevations of HFC. A correspondence of this kind is not *a priori* to be expected, since it is not known whether, and to what extent, a parallelism exists between the sensitivity of tissues to the effects of intracellular histamine generated by elevated HFC and the sensitivity to histamine acting from the outside.

It is generally stated that the rat has a low sensitivity to injected histamine. This statement requires modification. Rocha e Silva (1966), in a review article, refers to the experiments by Lecomte (1961), who in rats noted a distinct fall in blood pressure on intravenous injection of fractions of a microgram histamine per 100 g of body weight. This has also been observed in our laboratory (unpublished). Further, in rats anaesthetized with chloralose plus urethane we have recorded bronchoconstriction with 0.001 µg per rat intravenously (Fig. 55, facing p. 216). In our experience there is a very wide range in the sensitivity to injected histamine among individual anaesthetized rats. As far as there is a 'shock organ' in anaphylaxis in the rat, in our experiments manifest changes occurred in the small intestine which was filled with blood-stained fluid.

The presence of HFC in arteries (aorta), and its sharp elevation in anaphylaxis, is noteworthy. Histamine generated and acting in the vessel could cause damage to the walls of blood vessels, manifested as loss in tone and greatly increased permeability. The occurrence of exudate in the rat's small intestine is a case in point. Moreover, the observation that the HFC of arterial tissue can be greatly elevated should encourage further studies on intra-arterial histamine formation in physiological vasodilatation.

Mota's view on histamine release from disrupted mast cells in anaphylaxis has already been referred to. Histamine generated in embryonic and wound tissues, and in malignantly growing tissues, appears not to be associated with a high density of mast cells. In our study on anaphylaxis, the tissues were not examined for mast-cell density. Judged by published records on the number of mast cells in various tissues, it would appear that in some tissues considered to be poor in mast cells the observed elevation of HFC is on the same scale, or even greater, than in tissues known to be rich in mast cells. Histamine formed in anaphylaxis in the two species studied is, in our view, to a large extent of non-mast-cell origin.

The discovery that in reactions of hypersensitivity and anaphylaxis histamine is not merely released, but is also newly formed, places the involvement of histamine in a new light. Histamine formed at high rate, and acting intracellularly, may at least partly account for manifestations which formerly were not believed to be associated with histamine. The relative ineffectiveness of histamine antagonists in combating anaphylactic symptoms will now be better understood. The fact that the manifestations under discussion last much longer than the brief period of histamine release will now also be better understood. These reflections should in no way dim the realization that factors and circumstances as yet unrevealed contribute to the reactions of hypersensitivity and anaphylaxis.

It has been demonstrated that histamine is also released in human allergy and anaphylaxis (see Code, 1964, for references). There is little reason to doubt that in man, as in other species studied, histamine release is accompanied by elevations of HFC in various tissues including the lung. Lilja, Lindell and Saldeen (1960), in our laboratory, found that human lung tissue forms histamine, i.e. is endowed with HFC. The rate of formation *in vitro* was 2 to 15 nanogram ^{14}C-histamine per gram tissue in 3 hr. This comparatively low rate of formation is, however, of the same order of magnitude as that in the skin of normal rats. The finding of Lilja *et al.* should be considered in relation to the evidence presented by Schild, Hawkins, Mongar and Herxheimer (1951) that histamine apparently plays a role in the reaction of bronchial muscle of asthmatic subjects to allergens.

It should now be possible to demonstrate increased excretion of histamine in human allergy since non-isotopic methods for determining the principal histamine metabolites excreted in the urine in man have been developed, for methylhistamine by Fram and

Green (1965), and by White (1966), and for methylimidazole acetic acids (1-methyl-4-imidazoleacetic acid and 1-methyl-5-imidazoleacetic acid) by Granerus and Magnusson (1965), and Granerus (1968). In fact, this has been done in an instance of cold urticaria. In this particular case, not only the urinary excretion of histamine but also the excretion of two histamine metabolites was studied (Granerus, Svensson, Wetterqvist & White, 1969). The 28-year-old male patient was suffering from cold urticaria for about one year. In the actual experiment his whole body was bathed in increasingly cold water for 20 min on two successive days until general erythema and itching occurred. Urine was collected in 24 hr samples for a couple of days before, during and after exposure to cold. Histamine was determined on the guinea-pig ileum after purification on a cation-exchange column as described by Wetterqvist and White (1968). Methylhistamine was measured non-isotopically (Granerus, Wetterqvist & White, 1968) and methylimidazole acetic acid was determined non-isotopically (Granerus & Magnusson, 1965). It must be stated that it is not known with certainty whether methyl-5-imidazole acetic acid is a histamine metabolite. The results are given in Table 54. The most conspicuous results are the large increases in free histamine and methyl-4-imidazole acetic acid and the disappearance of methyl-5-imidazole acetic acid.

Table 54. The daily excretion of histamine, methylhistamine and the two isomers of methylimidazolacetic acid in a case of cold urticaria

Day	Urine volume (ml./24 hr)	Symptoms	Histamine base (μg/24 hr)	Methyl-histamine (μg/24 hr)	1-methyl-4-imida-zoleacetic acid (mg/24 hr)	1-methyl-5-imida-zoleacetic acid (mg/24 hr)
1	1580	No	18	175	1·5	2·8
2	2265	No	14	216	1·4	2·2
3	615	No	10	105	1·7	3·3
4	1570	Urticaria	80	361	8·3	0
5	1015	Urticaria	68	227	7·8	0
6	850	Slight urticaria	48	164	5·0	0
7	1550	No	16	182	1·6	0·5

(Granerus *et al.*, 1969, *Acta Allerg.* **24**, 258.)

It has been suggested by Mota (1966) that anaphylaxis in rats is due to an additive or synergistic effect between immediate and delayed hypersensitivities. The tuberculin reaction experimentally produced locally in the skin represents a particular kind of hypersensitivity reaction. In guinea-pigs Schayer and Ganley (1959) found that the HFC of skin excised 24 hours after the onset of a tuberculin reaction was about four times the control value. In rats the tuberculin reaction was examined by Graham and Schild (1967). They observed an increase in the HFC of skin with a maximum 12 hours after injection of antigen. The duration of the elevation was shorter than in our experiments on anaphylaxis. As we have explained earlier, elevation of HFC often coincides with a decrease in histamine content, and this occurred also in the experiments by Graham and Schild, as seen in Fig. 56.

The discovery that in anaphylaxis increased histamine formation

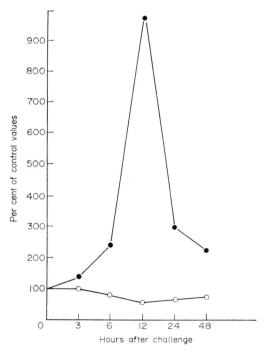

FIG. 56. Histamine-forming capacity (HFC) (●) and histamine content (○) of tuberculin skin reactions expressed as percentage of control. Each point is the mean of determinations on at least six animals (Graham & Schild, 1967, *Immunology* **12,** 725).

takes place in all tissues investigated in this respect, and that the elevated HFC persists for a long time after histamine release has ceased, appears helpful in explaining the failure of histamine antagonists to afford protection in the later stage of anaphylaxis. These drugs may relieve effects which occur initially during the brief period of histamine release. Histamine antagonists do not interfere with actions exerted by histamine formed and acting within tissue cells. Presuming that in anaphylaxis histamine is merely released, agents other than histamine are believed to account for the symptoms which are not alleviated by histamine antagonists. There are indications of such agents on record (for references until 1963, see Rocha e Silva, 1966). One member of this category of agents has now been recognized, namely the agent resulting from elevated tissue HFC.

15

POSSIBLE ROLE FOR HISTAMINE IN MAINTAINING HOMEOSTASIS

AMONG the mechanisms concerned in the regulation of homeostasis, principally the nervous and endocrine systems, histamine is rarely discussed as a conceivable associate. Nevertheless, observations and speculations have recently come forth which merit mention in this connexion.

Schayer's induced histamine. Schayer became impressed by the fact that a variety of circumstances and agents which induced increase in histidine decarboxylase activity are accompanied by changes in tone and permeability of terminal blood vessels. The correlation between changes in histidine decarboxylase activity and circulatory changes was taken by Schayer as a basis for the hypothesis that induced histamine serves as a governor of the functional state of the terminal vessels. The generalization assigning to induced histamine the role of adapting the microcirculation to the actual requirements of the tissues is drawn from circumstantial rather than direct evidence. Schayer's hypothesis regarding induced histamine and its control of the rate of blood flow through terminal vessels is likely to stimulate a direct experimental approach as soon as the inherent technical difficulties can be overcome. The topic of histamine as a factor in the humoral control of blood flow was further discussed in Chapter 10.

Histamine stimulates release of ACTH. Various conditions which impose a strain on homeostasis are apt to elevate tissue HFC. Owing to a rapid turnover rate in several tissues, histamine is continuously entering the blood stream, and if the concentration is sufficiently high, blood-borne histamine will excite the anterior pituitary gland to secrete ACTH. This has been shown to occur in the mouse, rabbit and dog. Histamine formed in the brain may also stimulate the secretion of ACTH, as set out in Chapter 13.

Elevation of HFC *in moribund animals.* In the ultimate strain on homeostasis, tissue HFC has been found elevated. In dying rats which had been subjected to trauma in a Noble–Collip drum, Schayer (1962*a*) found the histidine decarboxylase activity increased in lung and leg muscles, and in dying mice killed 42 hr after infection with *Klebsiella pneumonia,* Schayer (1962*b*) found this enzyme activity increased in all tissues investigated. In moribund ascites-tumour-bearing mice the HFC was substantially elevated, as seen in Table 55 (Kahlson, Rosengren & Steinhardt, 1963*a*). In the moribund state, the prevailing sympathetic nerve discharge is likely to excite the adrenal medulla. Catechol amines cause elevation of mouse tissue HFC (see Chapter 11). On the other hand, blood-borne histamine, resulting from elevated tissue HFC, is known to stimulate release of catechol amines, mainly that of adrenaline, in cats and dogs (Staszewska-Barczak & Vane, 1965).

Table 55. HFC in tissues from moribund cancer-inoculated and normal mice. The figures are counts/min per gram tissue. Values for normal tissues represent two to four determinations on pools from two to eight mice. Tissues were obtained from mice which were moribund on the 10th (A), 17th (B) and 20th (C) day after inoculation

	Normal mice	Mouse A	Mouse B	Mouse C
Lung	160–400	2100	8400	2900
Muscle	60–140	470	190	260
Spleen	600–1200	2400	2700	—
Kidney	430–17,000	430	360	630
Liver pH 7·4	70–90	590	1700	740
Liver pH 8, benzene added	105–140	290	740	330

(Kahlson *et al.,* 1963*a*, *J. Physiol.* **169,** 487.)

Regarding Table 55 it should be added that the enzyme activity generated in the tissues of moribund mice is identical with histidine decarboxylase according to the criteria by which this enzyme is characterized in our laboratory. Normal mice liver appears to contain some DOPA decarboxylase, since the activity is higher on adding benzene, but in the liver of the moribund mice benzene inhibited the enzyme activity, an inhibition which applies to histidine decarboxylase.

16

HISTAMINE METABOLISM IN GERMINATING SEEDS

THE germinating seed is a tissue in which growth is fast and easy to quantitate. This tissue can be kept in strictly defined environments. It appeared of interest to see whether the growing seed would generate HFC, and whether lowering of HFC by means of enzyme inhibitors would retard growth.

The occurrence of histamine as a natural product was first discovered in vegetable material, in ergot, by Barger and Dale, by Kutscher and by Ackermann, in 1910. The ensuing extensive inquiry about the pharmacological actions of histamine in animals left its association with plants largely unnoticed until 1948 when Werle and Raub demonstrated the presence of high concentrations of histamine in some plant species, particularly in spinach. The Munich workers showed that the histamine content of spinach seedlings increased during the course of germination and growth, and that the histamine which appeared during germination was newly formed. *In vitro*, the enzyme concerned decarboxylated L-histidine, and it lost activity completely on extraction or by the destruction of the cell structure by grinding. Semicarbazide in the concentration 0·001 M only irregularly retarded growth of spinach seed. Werle and Raub (1948) further demonstrated the presence of diamine oxidase in clover seedlings.

At about the same time Emmelin and Feldberg (1947) searched for a substance in the nettle hair which would release histamine from the human skin. They found histamine in the hair fluid in a concentration of 1 in 500 to 1 in 1000 and, in addition, acetylcholine in the hair fluid in a concentration of about 1 in 100 or stronger. Histamine and acetylcholine occur also in the leaf tissue of the nettle plant, but in weaker concentrations than in the hairs. These authors considered it possible that the agents are formed in the leaves, transported to the hairs and concentrated there.

Kahlson and Rosengren (1960) became impressed by experiments on germinating peas in the Department of Plant Physiology. We exposed germinating peas to semicarbazide. Figure 57 (facing p. 217), upper row, shows normal seedlings on the 10th day of germination, and the lower row, the seedlings exposed to semicarbazide 10^{-5} M. It should be mentioned that the germinating pea is rich in diamine oxidase, but the non-germinating pea contains none (Werle & Pechmann, 1949).

The investigations referred to above were carried further by Haartmann, Kahlson and Steinhardt (1966) whose experiments and observations will now be described. Spinach seeds (*Spinacia sleracea*) of high degree of uniformity and viability were used, nettles were collected in the surroundings of the Institute and the other seeds were obtained by the courtesy of the Institute of Plant Physiology, Lund. Before germination, the seeds were sterilized on the surface with 0·1% solutions of $HgCl_2$ and formalin to prevent fungal growth. The seeds were then soaked in 0·1% H_2O_2 to make them soft and turgid. The seeds were left to germinate in redistilled water at 24°C. The appearance of spinach seedlings at different stages of germination is shown in Fig. 58.

Determination of HFC. The tissues were finely cut with scissors and their HFC was measured isotopically. Only the spinach seed-

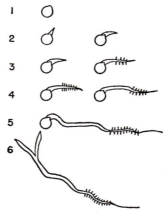

FIG. 58. Spinach seedlings at progressive stages of germination. The length of the root in mm is as shown: $1 = 0, 2 = 0 - 3, 3 = 3 - 9, 4 = 9 - 20, 5 = 20, 6 = 20 + $ cotyledons (Haartmann, *et al.*, 1966, *Life Sci*, Oxford **5**, 1).

lings exhibited a conspicuously high HFC among eleven species studied. The pH optimum of the histamine-forming enzyme was 5·0, at which pH all determinations listed in Table 56 were done. The rate of histamine formation was about 15% higher at 37°C than at 20°C and all incubations were carried out at 37°C. The rate of histamine formation increased steadily during germination to reach the highest value on the 6th day.

Table 56. HFC expressed as cpm/g of whole spinach seedlings during the stages of germination denoted by 1–6 in Fig. 58 (four series of experiments)

	Stage of germination					
Expt. no.	1	2	3	4	5	6
1	200	1400	1800	1300		
2	470	875	1600	1390		6700
3	625		1075	1550	2900	
4		852	675	2300	1400	4300
Mean	430	1042	1280	1650	2150	5500

(Haartmann, *et al.*, 1966, *Life Sci.*, *Oxford* **5**, 1.)

At stage 6 in Table 56 the various parts of the spinach seedlings were examined separately for HFC. In one pool the HFC was 4300 counts/min per gram and 5900 respectively in the root and shoot and the histamine content in the roots was 123 μg/g and in the shoots 395 μg/g. HFC was determined in various parts of germinating seeds of ten other plant species. In no instance were values higher than 25 counts/min per gram observed. The species examined were: *Pisum sativum*, *Linum usitatissimum*, *Helianthus annuus*, *Triticum*, *Hordeum*, *Avena sativa*, *Lycopersicum esculentum*, *Trifolium repens*, *Sinapis alba*, *Lepidium sativum*.

Histamine content. The last five listed species were examined for histamine content, and they yielded values from 0·3 to 1·1 μg/g. The spinach seedlings were richer in histamine, the content of which increased during the first six days of germination, as seen in Fig. 59.

Inhibition of histamine formation in vitro. Spinach seedlings at stages 5 and 6 were used. The actions of α-methylhistidine and α-methyl-DOPA at the concentrations indicated in Table 57 were

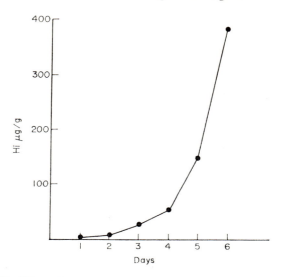

FIG. 59. Histamine content expressed in µg base/g of whole spinach seedlings during the first 6 days of germination (Haartmann *et al.*, 1966, *Life Sci.*, Oxford, **5**, 1).

investigated. It will be recalled that the former compound inhibits histidine decarboxylase, the latter inhibits enzymes of the DOPA decarboxylase class. Inhibition occurred with both compounds which indicates the presence of both enzymes. The concentrations required to produce significant inhibition appear high. However, the concentration of inhibitor existing within the tissue proper is unknown.

Table 57. Inhibitory action of α-methylhistidine and α-methyl-DOPA on the HFC (counts/min per gram) of 2 pools of whole spinach seedlings

Conc.	α-methylhistidine				α-methyl-DOPA	
	pool 1	inhib.	pool 2	inhib.	pool 2	inhib.
M	(count/min. per gram)	(%)	(count/min. per gram)	(%)	(count/min. per gram)	(%)
1×10^{-3}	3000	27	500	40		
5×10^{-4}			560	33	535	37
1×10^{-4}	3750	0	820	2	640	24
1×10^{-5}					635	25
0	3625		840		840	

(Haartmann *et al.*, 1966, *Life Sci.*, Oxford, **5**, 1.)

Inhibition of growth. Our observations on spinach seeds at stage 3 are presented in Table 58. In all experiments on seedlings treated with α-methylhistidine the length of the seedlings was less than that of the controls and the difference in length is statistically significant (t test, $P < 0.05$). Except in one case only the seedlings treated with α-methyl-DOPA grew as fast as the controls.

Table 58

Experiment on.	Controls	α-methyl-DOPA	α-mehid.
1	6.54 ± 0.14 (95)	5.65 ± 0.26 (33)	5.69 ± 0.03 (96)
2	6.67 ± 0.39 (27)	—	5.57 ± 0.30 (31)
3	6.11 ± 0.63 (23)	6.90 ± 0.52 (22)	4.71 ± 0.40 (18)
4	5.40 ± 0.35 (22)	—	4.59 ± 0.30 (23)
5	5.45 ± 0.19 (37)	5.23 ± 0.31 (21)	4.88 ± 0.22 (28)
Total	6.19 ± 0.12 (204)	5.89 ± 0.20 (76)	5.30 ± 0.09 (196)

(Haartmann *et al.*, 1966, *Life Sci.*, Oxford **5**, 1.)

Comment: The experiments on germinating spinach seeds, but not on the others, revealed similarities to the histamine metabolism in mammalian tissues. The histidine decarboxylase activity increased with the growth of the seedlings, here too this particular enzyme activity was inhibited by α-methylhistidine and the inhibition of this enzyme retarded growth. Beside this enzyme a different one was present which apparently was a member of the DOPA decarboxylase class since it was inhibited by α-methyl-DOPA. Inhibition of this latter enzyme did not retard growth. In order to obtain inhibition of growth rather high concentrations of the inhibitors, 10^{-3} M solutions, were employed. This high concentration of α-methyl-DOPA, however, did not retard the growth of the seedlings. It should be taken into consideration that in the *in vivo* experiments the actual concentration of the inhibitors inside the tissue was unknown.

17

HISTAMINE FORMATION IN PREGNANCY

AFTER the disclosure that in the female rat, under specified conditions, the urinary excretion of free histamine reflects the rate of endogenous histamine formation, the discovery that the rat produces excessive amounts of histamine during a certain phase of pregnancy lay at the door. The elevated HFC was found to reside in the rapidly growing foetus. This finding, in turn, sowed the seed from which grew the idea of an association between high HFC and rapid tissue growth. In our early work in this field we depended solely on the non-isotopic *in vivo* method for whole-body HFC determinations.

Rat

The mean 24-hr histamine excretion is steady before pregnancy, and no striking changes occur during the two first weeks of gestation. However, on the 15th day of pregnancy a distinct and steep increase in histamine excretion occurs, a typical instance of which is seen in Fig. 60. The increase is maintained until a day or two before the birth of the young. On the day before term the urinary histamine falls precipitously towards the pre-pregnant level. Aminoguanidine, which inhibits histaminase, elevates the histamine excretion 3–4 times, but does not alter the pattern or time-course of excretion and does not interfere with the course of pregnancy or parturition in rats (Kahlson, Rosengren & Westling, 1958; Banik, Kobayashi & Ketchel, 1963).

In the attempt to identify the tissue(s) responsible for the increased formation of histamine during the last third of pregnancy, guidance was provided by the observation that the urinary excretion of histamine was proportional to the number of young in the litter. It thus appeared logical to study the effect of the removal

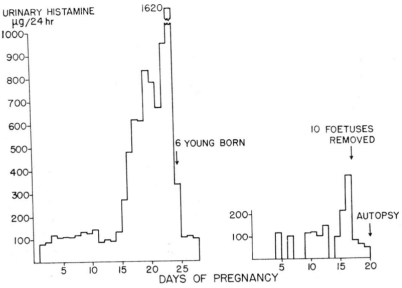

FIG. 60. Urinary excretion of histamine in undisturbed pregnancy (left side of the figure) and in a rat where the foetuses were removed at the 17th day of pregnancy (right side). Throughout the whole course of the observations the rats were under the influence of the histaminase inhibitor aminoguanidine (Kahlson, Rosengren, Westling & White, 1958, *J. Physiol.* **144**, 337).

of the foetuses on the urinary excretion of the amine. It was known from various reports that if the foetuses are removed deliberately without dislodging the placentae, the rat remains physiologically pregnant by a series of criteria easily recognizable at autopsy (see Newton, 1949, for references). This fact was utilized by removing the foetuses at the 17th and 19th day of pregnancy, i.e. after the onset of demonstrable increase in the excretion of histamine and before the increase had reached the usual peak value. On the removal of the foetuses, urinary histamine fell sharply and rapidly reverted to the pre-pregnant level or even lower, as in Fig. 60. This experiment indicates that the increased formation of histamine during the last third of pregnancy depends on the presence of the foetuses and not on other structures characteristic of pregnancy.

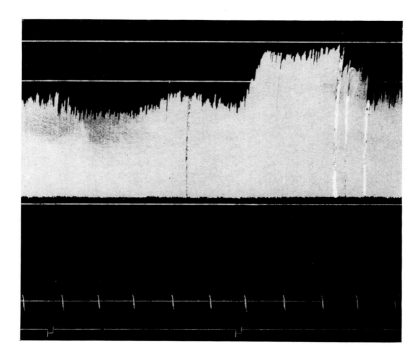

F<small>IG</small>. 55. Rat, 276 g, chloralose-urethane anaesthesia. From below upwards: injection signal, time marks in minutes and bronchial tone registered by the method of Konzett & Rössler (*Arch. exp. Path. Pharm.*, 1940, **195**, 71). At the first and second signal marks 1·0 ml. saline and 0·001 μg histamine, respectively, were injected intravenously (Svensson *et al.*, unpublished).

[*facing p. 216*]

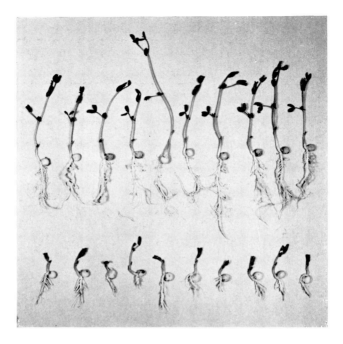

FIG. 57. (Kahlson & Rosengren, 1960, *J. Physiol.* **154,** 2P).

[*facing p. 217*]

HFC *determined* in vitro *in the foetus and in tissues from pregnant and non-pregnant rats.* Various tissues of the pregnant mother were examined at times when the urinary excretion of histamine was high. Foetuses were removed and examined during the period of high histamine excretion and during the period of decreasing excretion. The histamine formation was also estimated in newborn young. The results are summarized in Table 59 from which it will be seen that of the tissues examined in pregnant as well as in non-pregnant rats the stomach is the most potent in the formation of histamine, followed next by the lungs, whereas the uterus, the maternal and foetal placentae and the other tissues examined show low activities. In the adult rat tissues investigated we did not find any conspicuous difference between pregnant and non-pregnant female rats as regards histamine formation. By contrast, the foetuses were very potent in producing histamine during the period of increased urinary excretion of histamine.

We made no attempt to examine foetal HFC during the first two weeks of gestation. At the 9th day of gestation the embryo weighs 0·006 g, at the 11th day 0·08 g and at the 14th day 0·27 g (Misrahy, 1946). At the 15th day the single foetus has an average weight of 0·35 g, increasing within the short period of a week to about 5 g at term. The 15th day, at which increased histamine excretion is first manifest, marks the beginning of a large daily gain in weight of the foetuses which, from the 15th to the 16th day, increase in weight by about as much as during the previous 2 weeks.

Distribution of HFC *in rat foetuses, new-born, and adults.* In this study, Kahlson, Rosengren and White (1960) determined HFC in a variety of foetal tissues on the 19th, 20th and 21st day of gestation. Histamine excretion by the mother is at the peak on the 19th or 20th day of gestation, and tends to fall on the 21st day. It will be seen from Table 60 that the foetal liver produces histamine at an enormously high rate with a HFC about 100 times higher than in the rest of the foetal body. A mast-cell tumour of a dog investigated in our laboratory (Lindell, Rorsman & Westling, 1959) was about 100 times less active per unit weight in forming histamine than the foetal liver. All foetal tissues investigated are potent producers of histamine. From values obtained in this study it was calculated that the liver accounts for about 80% of the histamine formed by the foetus. Mast cells were not found in the

H

Table 59. Histamine formation in rat tissues, foetuses and new-born young. The figures are counts/min per gram tissue

	Non-pregnant females			Mothers	
Lung	89	—*	—	90	150
Stomach†	—	690	450	980	370
Kidney	16	3	—	<4	3
Uterus	<44	5	—	<7	6
Maternal placenta	—	—	—	<10	—
Foetal placenta	—	—	—	<5	9
Brain, total	26	20	35	—	—
Abdominal skin	—	—	—	—	9

	Foetuses Day of pregnancy			New born	
1	17	1110	1	<3 hr old	76
2	17	2470	2	<1 hr old	76
3	17	2360		Collected	{210‡
4	19	1510	3	from the	{400
5	20	2480		vagina	
6	21	{ 290‡	4	during	{460‡
		{ 380		parturition	{970
7	22	410§			
8	23	520			
9	23	{1080‖			
		{1090			

*—Not determined.   † Total stomach wall.   ‡ Different foetuses.
§ Two foetuses pooled.   ‖ Duplicates.
(Kahlson, Rosengren, Westling & White, 1958, *J. Physiol.* **144**, 337.)

foetal tissues examined except for the skin. On the day before term, when the maternal excretion of histamine is on the wane, the activity of the foetal liver has fallen to the low level found in the new-born, and within 2–3 days the histidine decarboxylase activity of the liver regresses to the low level found in the adult animal. Whereas the time of the first appearance of high histidine decarboxylase activity in the foetus has not been established, the time of its disappearance in the foetus is well defined.

Histamine content of foetal tissues. Since the published figures for the histamine content of the rat foetus are inconsistent (for references see Kahlson, Rosengren & White, 1960), and because

Table 60. Rate of histamine formation in 3 hr in rat foetuses, new-born, and adults, expressed as counts/min per gram tissue. 1 µg ^{14}C-histamine gives 600 counts/min

	Foetus 19 days	Foetus 20 days	Foetus 21 days	New-born < 3 hr	Young 2–8 days	Adult pregnant
Whole body	—	3440	340	860	15	—
Liver	14,000 (23,000)*	8730	2280	2240	5–40	5
Stomach	2800	—	—	—	—	370–980†
Lungs	1900	930	150	60	5	90–150†
Heart	—	—	140	30	0	—
Kidneys	—	—	190	—	0	< 5†
Stomach + intestines	—	1250	150	—	—	—
Brain	10	5	20	5	0	—
Rest of the body (muscle, skin, bone)	130 (220)*	360	70	60	30	—

* Figures in brackets are the means of determinations in three litters; the other figures in this column are from a single litter. All the results in the second column are from one litter and so also those in the third and fourth.

† Figures quoted from Kahlson, Rosengren, Westling & White, 1958.

(Kahlson, Rosengren & White, 1960, *J. Physiol.* **151**, 131.)

our rats were fed on a special histamine-free diet, it was necessary to study the histamine content of the foetus anew. We examined two litters on the 19th day, and one on the 21st day of pregnancy, one at 6 hr, one at 8 hr and one at 8 days after birth, pooling the livers, lungs, etc., from the members of each litter. It will be recalled that at the 19th day the HFC of the foetus is about maximal, whilst the 21st is the day of steep regression in the rate of histamine formation. The values obtained for histamine content are shown in Table 61. At the time of maximum histamine production in the whole foetus, the average total concentration in the body is not greater than a few micrograms per gram tissue. It is evident that the newly formed histamine is not bound in the foetus but is transferred rapidly to the mother, where it is partly catabolized and partly excreted. The relatively high histamine content of the foetal liver should not be taken as evidence of its possession of binding sites for histamine, but rather it indicates the limited removal rate of histamine so rapidly formed.

Binding of histamine formed in the foetus. To see what proportion of newly formed histamine remains bound, in the sense that it cannot be easily washed away, determinations were done *in vitro* on

Table 61. Histamine content of tissues of rat foetuses and new-born
young. The figures denote μg histamine/g tissue

	Whole foetus	Foetus without liver	Liver	Lung
Foetus, 19th day of pregnancy	—	2·3	18·6	—
Foetus, 19th day of pregnancy	2·1	1·4	23·5	0·7
Foetus, 21st day of pregnancy	1·3	1·3	4·6	0·7
Young, 6 hr after birth	—	8·5	1·0	1·0
Young, 8 hr after birth	—	10·1	0·9	0·8
Young, 8 days after birth	—	11·7	0·8	—
Adult females (Gustafsson *et al.* (1957)	5·1–6·9	—	—	3·3–8·2

(Kahlson, Rosengren, & White, 1960, *J. Physiol.* **151**, 131.)

whole foetuses, liver, lung and skin. The minced tissue was
incubated with ^{14}C-histidine as described in Chapter 5. After
3-hr incubation the mixture was centrifuged at 1000 *g* for 10 min.
The supernatant fluid was decanted and the residue was washed
with buffer solution and centifuged twice. The ^{14}C-histamine
content of the tissue mince and the combined supernatants were
determined separately. The results are presented in Table 62,
which shows that in the whole foetus about 12% of the newly
formed histamine was retained, in the liver 20%, in the lung 40%
and in the skin 45%. This experiment shows that the histamine
formed in foetal tissues easily diffuses into the surrounding fluid
and that, except for the lung and skin, only a small proportion of it
is retained in the tissues. In the rat foetus mast cells have been
found only in the skin (for references, see Dixon, 1959). Foetal
histamine is predominantly of non-mast-cell origin. It is noteworthy that the foetal lung, like that of the adult, is provided with
a non-mast-cell arrangement capable of retaining histamine, as in
the foetal and adult skin. In an experiment, similar in type to ours,
Schayer, Davis and Smiley (1955) found that in the skin of the
adult rat 2/3 of the histamine formed was retained in the tissues,
whilst 1/3 was in the supernant.

Physiological significance of high foetal HFC *in the rat.* Smooth
muscle, including that of the uterus, contracts when adequately
distended. The distension of the uterus by its growing contents
would lead to contractions and premature birth if a strong in-

Table 62. Binding of newly formed histamine in rat foetuses and in foetal tissues expressed as counts/min per gram tissue. A high ratio in the last column indicates a low histamine-binding capacity and vice versa

Tissue	Day of pregnancy	Super-natant (a)	Tissue (b)	Ratio (a)/(b)
Foetus (3) ⎫ * same litter	17	2100	350	6·0
Foetus (4) ⎭	17	2100	310	6·8
Foetus (2) ⎫ same litter	17	2000	220	9·1
Foetus (2) ⎭	17	2200	270	8·1
Foetal liver	20	3490	780	4·5
Foetal lungs	20	210	150	1·4
Foetal skin	20	140	120	1·2

* The figures in parentheses denote the number of foetuses pooled in one sample.

(Kahlson, Rosengren, & White, 1960, *J. Physiol.* **151**, 131.)

hibition of some kind did not exist. The nature of this inhibitory influence, which ceases at the time of parturition, is unknown. It has been known for a long time that histamine inhibits the rat uterus smooth muscle (Guggenheim, 1912). In the rat, the foetal histamine formation, both in magnitude and time-course, appears designed to contribute to establishing the inhibitory action required to prevent premature uterine contraction, as pointed out by Kahlson, Rosengren and Westling (1958). This interpretation has been substantiated by unpublished observations in our laboratory on the postponement of parturition by injecting 'long-acting' histamine, and by Wiskont-Buczkowska (1961), who recorded prolongation of pregnancy for two days by repeated daily injections of histamine. However, the fact that in the rat, after removal of the foetuses, the placentae are delivered at the end of the normal period of pregnancy, appears difficult to reconcile with the idea of foetal histamine formation as a major factor operating to maintain the uterine muscle quiescent until the day of term. Nevertheless, foetal histamine may play a role as a contributing, although not imperative, factor.

The discovery of high histidine decarboxylase activity in the rat foetus marks the starting-point for investigations on an association between histamine formation and tissue growth. If foetal HFC and the formation thereby of 'nascent histamine' were essential to the growth and metabolism of the foetus, then its suppression would be expected to interfere with foetal development. In fact, on suppressing foetal histamine formation, arrest of growth and consequent death of the foetal ensues. In rats fed on a pyridoxine-deficient diet, we administered semicarbazide for 8 days during various phases of pregnancy, from the time of implantation of the ovum onward. On examining the uterine contents, the foetuses were found dead and their growth seemed to have been arrested at the very day when inhibition of histamine formation was greatest, about 85% inhibition. In the typical instance shown in Fig. 61 (facing p. 248), at the 19th day of pregnancy (when the weight of a normal foetus is about 2 g), the foetuses were dead and reduced to a tiny mass of disintegrated material; the placentae were small and in a state of regression (Kahlson & Rosengren, 1959).

As discussed in Chapter 8, semicarbazide inhibits enzymes other than histidine decarboxylase; such enzymes include various amino acid decarboxylases and the diamine oxidases, including histaminase. However, except for histamine, no other substance whose enzymatic formation is known to be inhibited by semicarbazide has so far been found to be specifically related to foetal development. The formation of 5-hydroxytryptamine is inhibited by the inhibition procedure employed in this particular experiment, but this compound is reported not to be produced in increased amounts in the rat during pregnancy (Angervall, Enerbäck & Westling, 1960). Regarding histaminase, we have found that its complete inhibition by aminoguanidine in the rat has no effect on the course of pregnancy or the fitness of the new-born (Kahlson, Rosengren & Westling, 1958) and, in addition to confirming this result, Banik, Kobayashi and Ketchel (1963) found that aminoguanidine had no apparent effect on the implantation of blastocysts in the rat.

The specific and non-toxic inhibitor of histidine decarboxylase, α-methylhistidine, was not available to us in the quantity required for this particular experiment which is likely to be repeated, when once a specific in vivo inhibitor of histamine formation becomes available. We are aware that in the state of pyridoxal deficiency disturbances of anabolic processes exist which perhaps are more damaging to the foetus than to the mother, e.g. interference in transaminase reactions in which pyridoxal phosphate acts as an amino-carrier.

Mouse

The observations in the pregnant rat incited studies of histamine formation during pregnancy in other species also. Hence it was noted that the non-pregnant mouse fed on a partly synthetic histamine-free diet excreted free histamine in the urine. After mating the histamine excretion increased steeply from the very first day of pregnancy onwards (Rosengren, 1963). This was at variance with the situation in the rat. As pregnancy proceeds histamine excretion rises to values sometimes 100 times the pre-pregnant. After delivery the urinary histamine falls, at first steeply, then more slowly, the pre-pregnant level not being attained until a few weeks after delivery. This, again, is at variance with the situation in the rat. The events in the mouse, which in their time-course are distinct from those in other species, are illustrated in Fig. 62.

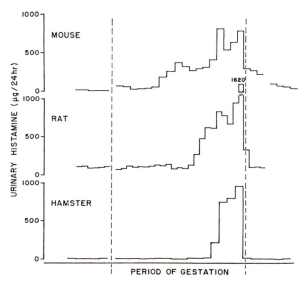

FIG. 62. Daily urinary excretion of free histamine during pregnancy in the mouse, rat and hamster. The periods of gestation were 19, 24 and 16 days, respectively. The rat was under the influence of aminoguanidine. Note that in all species the highest values are obtained in the last third or fourth of pregnancy. Note also that the mouse differs from the other two species by showing an early increase in histamine excretion and persistence of the increase after delivery (Kahlson & Rosengren, 1968, *Physiol. Rev.* **48**, 155).

The results obtained in the mouse *in vivo*, non-isotopically, could be corroborated isotopically in that after injection of ^{14}C-histidine in pregnant mice larger amounts of ^{14}C-histamine and ^{14}C-methylhistamine were excreted than in non-pregnant mice.

The mouse foetus, like that of the rat, forms histamine at a high rate (Table 63). The histidine decarboxylase activity of the whole foetus rises with increasing age of the foetus up to term, whereupon the level of enzyme activity falls. There are, however, great differences between mouse and rat in the regional distribution of foetal histidine decarboxylase. In the mouse foetus the HFC resides mainly in the skin, whereas in the rat foetus the liver constitutes the principal site of histamine formation.

Table 63. Histamine-forming capacity (HFC) of mouse foetal tissues (counts/ min per gram tissue)

	Whole foetus	Skin	Brain	Thoracic +abdom. viscera	'Rest of the body'
Foetus					
18th day of pregnancy	73,000	220,000	1400	8000	39,000
19th day of pregnancy	91,000	203,000	760	5400	48,000
Young					
< 18 hr old	14,000	24,000	410	840	13,000
About 36 hr old	4000	10,000	330	3700	2800
7 days old	970	1600	70	1100	340

(Rosengren, 1963, *J. Physiol.* **169**, 499.)

A further pertinent species difference of significance is that a conspicuous change occurs in the kidney HFC of the pregnant mouse. As early as the first day of pregnancy the kidney HFC attains very high levels that persist during the entire course of pregnancy and remain elevated even for a few weeks after delivery (Rosengren, 1963). The mouse kidney enzyme appears to be identical with histidine decarboxylase, as it is strongly inhibited by α-methyl histidine and only weakly inhibited by α-methyl-DOPA (Rosengren, 1962). It will be shown in Chapter 18 that the kidney of the non-pregnant mouse is singularly useful in exploring agencies and circumstances conducive to elevation of histidine decarboxylase activity. In this connexion it should be mentioned that the rat kidney HFC is very low and that the kidneys of the guinea-pig and rabbit form histamine *in vitro* by the agency of DOPA decarboxylase.

Hamster

The female hamster, like the rat and mouse, excretes free histamine in the urine. The gestation period in this species is short, 16 days. Marked deviations from the pre-pregnant level of histamine excretion do not occur during the first two-thirds of the gestation period. However, on the 11th day of pregnancy histamine excretion increases and rises to very high values during the three days prior to term. On the day of parturition a steep fall in the amine excretion occurs. From the first day after delivery onwards the excretion is maintained at about the pre-pregnant level (Fig. 62) (Rosengren, 1965).

Following the subcutaneous injection of ^{14}C-histidine, much larger amounts of ^{14}C-histamine and ^{14}C-methylhistamine were excreted in the pregnant animal than in non-pregnancy.

In attempts to identify the site(s) of high HFC in pregnancy, various tissues were excised and examined for their histidine decarboxylase activity *in vitro*. A definite HFC was observed in the lung of the pregnant animals during the days before term, whereas none was detectable in this tissue from non-pregnant animals. However, none of the eight extra-uterine tissues examined (Table 64) exhibited in pregnancy an elevated HFC which could reasonably account for the excessive amounts of urinary histamine.

Further exploration revealed that the large amount of histamine

Table 64. HFC of various tissues of adult female hamsters. Figures represent counts/min per gram

Day of pregnancy	Abd. skin	Lung	Liver	Spleen	Gastric mucosa	Small intest	Kidney	Skeletal muscle
11	20	0*	0	40	5000	150	2100	0
14	80	90	50	100	3100	90	1900	20
	150	100	100	90	2200	—†	1300	—
15	80	30	0	30	4500	—	1200	—
	50	150	—	90	1400	170	1200	110
16	80	210	40	170	3700	210	1000	0
Non-pregnant								
	100	0	70	70	9000	140	1000	30
	40	0	80	30	7300	100	810	0
	120	0	50	40	17000	130	940	10
	110	0	10	60	1100	—	1200	—

* 0 not detectable.   †—not determined.
(Rosengren, 1965, *Proc. Soc. exp. Biol.*, N.Y. **118**, 884.)

excreted originates from the uterus, predominantly in the placenta (Table 65). The uterine wall and the foetal membranes (placenta removed) also showed a rather high HFC.

Table 65. HFC (counts/min per gram) of uterus and its contents

Day of pregnancy	Uterus wall	Placenta	Foetal membranes (placenta removed)	Whole foetus
11	2900	120,000	1600	0*
	150	48,000	—†	—
14	800	820,000	900	100
	560	1,100,000	2500	260
	250	860,000	1200	60
	—	920,000	—	—
	—	930,000	—	—
15	4400	830,000	3800	—
	1900	950,000	—	270
	250	1,000,000	1300	70
	—	1,100,000	—	170
16	160	970,000	400	100
	150	580,000	630	40
Non-pregnant	40			
	30			
	10			

* 0 not detectable.   †—not determined.
(Rosengren, 1965, *Proc. Soc. exp. Biol., N.Y.* **118**, 884.)

To investigate whether histidine decarboxylase was uniformly distributed in the placenta or was confined to some particular layer(s), Rosengren (1965) separated the placenta of five hamsters into 7 to 10 layers, each consisting of 10 sections of 30 μ thickness. Pooled sections from each layer were determined for HFC and the histological structure was studied. The values for enzyme levels are detailed in Fig. 63, in which the histidine decarboxylase activity of the 10 layers of the placenta from hamster No. 5 (Table 66) is represented in relation to a microphotograph of a transverse section through these layers. The enzyme is preferentially distributed, with peak activity in layers II, III and IV. These layers are distinguished by the presence of particular kinds of cell, among them giant cells which cytologically have been described by Ward-Orsini (1954).

Table 66. Vertical distribution of placental HFC (counts/min per gram) in 5 hamsters on 14th (No. 1-3) and 15th Days (No. 4-5) of pregnancy. Roman figures refer to number of the layers as described in text. Layer I is the one facing the uterine wall

| | Hamster no. | | | | |
	1	2	3	4	5
I	390,000	920,000	1,300,000	810,000	590,000
II	1,400,000	1,900,000	1,900,000	1,700,000	1,400,000
III	1,900,000	1,100,000	1,300,000	1,600,000	1,900,000
IV	1,600,000	430,000	400,000	1,700,000	2,100,000
V	870,000	140,000	85,000	730,000	1,100,000
VI	460,000	62,000	54,000	120,000	250,000
VII	220,000	10,000	41,000	21,000	100,000
VIII	85,000			—*	
IX	72,000				21,000
X					

* Lost.

(Rosengren, 1965, *Proc. Soc. exp. Biol.*, *N.Y.* **118**, 884.)

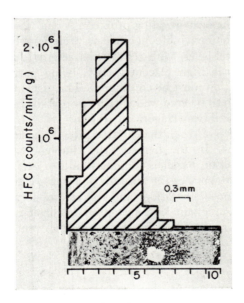

FIG. 63. HFC profile and microphotograph of hamster placenta on 15th day of pregnancy (Rosengren, 1965, *Proc. Soc. exp. Biol. N.Y.* **118,** 884).

The non-pregnant hamster, like all mammalian species studied in this laboratory, forms histamine, and in this animal the gastric mucosa has the highest HFC. The principal pathways of histamine inactivation in the hamster are methylation and oxidative deamination. In this respect the hamster is similar to the mouse. There is no indication that an alteration in the general pattern of histamine catabolism occurs in the mouse or hamster during pregnancy (Rosengren, 1963, 1965).

There is no considerable difference in the tissue HFC outside the uterus between the pregnant and non-pregnant hamster, except for the lung. This is in contrast to the findings in the rat and mouse, in which the gastric mucosal HFC is elevated in pregnancy (see Chapter 9). The latter species shows, in addition, a very high elevation in kidney HFC during gestation. These species differences, like others to be discussed later, cannot be accounted for at the present.

The striking height of hamster placental HFC which by far exceeds the levels observed in any normal tissue so far studied presents a particularly puzzling problem. A clue has been found in the fact that the enzyme concerned is preferentially localized in definite placental layers with cells of characteristic structure. The possibility of the high placental HFC being part of a specific metabolic function must be considered. The placenta of mammals has been shown to be endowed with an impressively wide range of enzymatic capabilities (Hagerman, 1964).

Also the time-course of the elevated rate of histamine formation poses problems. In the hamster the time-course of increased histamine excretion parallels the one found to occur in the pregnant rat; in both species the onset of detectable elevation of HFC is at a late stage of gestation, and the rate of formation falls to about the non-pregnant level just prior to, or at term. This fact may suggest involvement in the complex events involved in parturition.

Man

Only a few studies of histamine formation in human pregnancy are so far on record. With the foetus (weight 90–285 g) *in utero* we collected blood from the umbilical artery and vein and, further, from a carotid artery after the foetus had been removed. In foetal carotid artery and umbilical artery plasma histamine was present in easily demonstrable amounts, whereas maternal plasma, like human plasma generally, was devoid of histamine in measurable concentrations. Also, there was more histamine in the umbilical artery plasma than in the umbilical vein plasma (Kahlson, Rosengren & White, 1959). Similar results have been reported by Bjurö, Lindberg and Westling (1961). On the other hand, Mitchell and Cass (1959) found no significant difference in the histamine content of foetal and maternal blood. Moreover, Mitchell (1963), employing a method similar to Waton's (with or without benzene), arrived at the conclusion that human foetal histidine decarboxylase displays properties—activity greatly increased by benzene—which should now be ascribed to DOPA decarboxylase.

More direct evidence of histamine formation in human foetal tissue was obtained by Lindberg, Lindell and Westling (1963*b*) who employed an isotopic technique which indisputably measures histidine decarboxylase. They found rather low levels of activity; the activities in order of magnitude were: spleen, hypothalamus, stomach, abdominal skin and liver.

The literature pertinent to histamine metabolism in man, up to December 1961, has been reviewed by Lindell and Westling (1966), and further work in this field has been discussed in the chapter on histaminase (Chapter 7).

18

HISTAMINE FORMATION AND HORMONES

A CONNEXION between histamine metabolism and hormones has been evident for more than thirty years from the discovery that in human pregnancy the plasma exhibits strong histamine destroying activity. This lead has been of little avail to students of the physiology of histamine. The finding of elevated HFC in foetuses attracted attention to the possibility of a relation between histamine formation and alterations in the hormonal status prevailing during pregnancy. Accordingly, it seemed appropriate to begin the experimental analysis by investigating the effects of sex hormones on histamine metabolism.

Sex differences and miscellaneous. Direct indications of a relationship between hormones and histamine metabolism are on record. In the study of the biogenesis of tissue histamine in germ-free rats, a sex difference was demonstrated in that in the male the urinary histamine was largely in a conjugated form, whereas the female excreted histamine predominantly in the free form. A sex difference exists also in ordinary rats (Leitch, Debley & Haley, 1956) and is presumably due to a more efficient inactivation of histamine in the male in which methylation is the principal catabolic pathway (Westling, 1958). Testosterone administration increases the proportion of methylhistamine in the urine of castrated male and female rats (Westling & Wetterqvist, 1962). A peculiar sex difference in histamine metabolism, the nature of which is obscure, exists in mice, in which the kidney of the female is endowed with a much higher HFC than that of the male (Rosengren & Steinhardt, 1961). Here, it would appear, is a target organ specifically responsive to some unknown circumstance conducive to raising histidine decarboxylase activity.

Treatment of rats with thyroid hormones (thyroxine or lio-thyronine) has been shown by Bjurö, Westling and Wetterqvist (1961, 1964*a*) to result in increased rate of histamine formation, evident from the greater amounts of ^{14}C-histamine excreted after injection of ^{14}C-histidine.

The influence of glucocorticoids on histamine formation has been extensively investigated and currently reviewed by Schayer (1967). In the rat these compounds reduce the rate of histamine synthesis in all tissues studied except the stomach, which shows an increased rate of histamine formation.

Oestrogenic compounds. Removal of the foetuses in the rat, without otherwise interfering with the state of pregnancy, is followed by a sharp fall in whole-body histamine formation to about the pre-pregnant level (Fig. 60). The operation was performed without detaching the placentae, which were later delivered at the time of normal parturition. It is generally believed, although not proved, that in this situation the placentae continue to produce hormones in much the same manner as during normal pregnancy. If placental and ovarian hormones are instrumental in raising the histidine decarboxylase activity, then in the rat the embryonic tissues are the principal target organs. In the rat another target organ exists, the responsiveness of which appears rather dubious in character, the gastric mucosa, the HFC of which is elevated in pregnancy, as discussed in Chapter 9.

In the pregnant mouse, not only the foetuses, but also the kidneys of the mother display greatly elevated HFC. The mouse kidney therefore served as an excellent target organ for testing the effects of oestrogens and progestins on the kidney HFC in the non-pregnant female. In the actual experiments in mice, administration of progesterone caused no significant change in whole-body histamine formation, as expressed by urinary free histamine. By contrast, oestrogens administered as oestradiol produced a substantial increase in urinary histamine. These results are illustrated in Fig. 64, which also includes a control given saline only.

In order to identify the site(s) of elevated histidine decarboxy-lase activity following administration of oestradiol, ovariectomized mice were given 100 μg oestradiol monobenzoate daily, on two successive days. Various tissues were excised three weeks after the

MAGDALEN COLLEGE LIBRARY

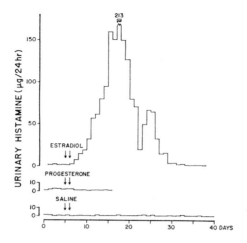

FIG. 64. Daily urinary excretion of free histamine in ovariectomized mice. Oestradiol, 10 μg daily intraperitoneally, on 2 successive days, is followed by a long-lasting increase in urinary histamine excretion. Progesterone, 1 mg daily subcutaneously, on 2 successive days, does not detectably alter histamine excretion. The same applies to intraperitoneal injections of 0·9% NaCl solution (Rosengren, 1966 *Acta Univ. Lundensis*, Sec. II, no. **8**, 1).

onset of treatment, i.e. around the day when histamine excretion was known to be at the peak level. A striking elevation of kidney HFC occurred in each animal, 10 to 50-fold, in agreement with the high rate of histamine formation noted in the kidney of the pregnant mouse. In the other tissues investigated (Table 67) there was no change of the magnitude seen in the kidney (Rosengren, 1966). The results from the studies on rapid growth and protein synthesis, described in later chapters, should not overshadow the occurrence of a strikingly elevated HFC in the kidney of the pregnant mouse. Here, obviously, the elevated HFC is not associated with the process of growth in a way similar to that in the other instances recorded. The high kidney HFC may perhaps be functionally related to the high HFC of the hamster placenta.

It has been suggested by Szego and her colleagues that histamine may serve as a mediator of oestrogen action on the uterus. Following the injection of 0·54 mg histamine in the uterine horn

Table 67. Effect of oestrogen treatment on histidine decarboxylase activity in tissues of ovariectomized mice. The figures are ng ^{14}C-histamine formed/g

	Controls	Oestradiol		Controls	Oestradio
Skin	0*	40	Gastric	590	830
	40	30	mucosa	850	760
	100	90		160	360
	—†	—		—	—
	50	—		—	—
		100			—
Lung	60	190	Small	0	0
	130	190	intestine	0	0
	30	110		0	0
	—	—		—	—
	—	—		—	—
		—			—
Liver	10	30	Kidney	1900	9200
	10	20		120	8400
	10	20		30	5500
	—	—		1400	13,000
	10	—		440	19,000
		20			28,000
Spleen	150	530	Uterus	70	—
	190	240		100	0
	100	180		260	30
	—	—		—	—
	—	—		0	—
		—			60

* 0 not detectable. †—not determined.

(Rosengren, 1966, *Acta Univ. Lundensis*, Sec. II, no. 8, 1.)

in ovariectomized rats they found that the incorporation of ^{14}C originating from glycine and lysine into selected metabolites was augmented in uterine segments *in vitro* at a rate similar to that induced by oestrogen. These investigators suggested that oestrogen-responsive metabolic pathways are capable also of stimulation by histamine in the absence of the hormone. If this could be proved, it would stand as a great contribution towards the understanding of the mode of the metabolic functions of histamine. Following local instillation of histamine, mitotic activity, especially in the luminal epithelium and endometrial glands, is reported to be stimulated. Moreover, injection of oestrogen is reported to release histamine in the uterus. These experiments have been reviewed by Szego (1965). A less optimistic view of this topic is taken by Martin (1962) who from the available evidence finds it doubtful that histamine release is involved in responses of target organs to oestrogens and, further, excludes histamine itself as an agent exerting oestrogenic action. In concluding this section, reference should be made to the suggestion set forth in Chapter 13, that blood-borne histamine may serve as a potential stimulant for the pituitary-adrenal system.

19

WOUND HEALING

HAVING disclosed the association between foetal growth and high HFC, it appeared essential to study different types of rapid growth. We therefore turned to investigate the reparative growth in healing skin wounds. This choice appeared particularly attractive because in the skin the rate of histamine formation can be lowered or raised experimentally. Experiments on healing wounds in our laboratory have so far been carried out only in the rat and human (Kahlson, Nilsson, Rosengren & Zederfeldt, 1960; Kahlson, Rosengren & Steinhardt, 1963b).

Wounds. Female rats were used, and two wounds 6 cm long were inflicted in the skin of the back. The incisions were made through the shaved skin and the subcutaneous muscle layer, and the wounds were closed by silk suture stitches. To prevent the rats biting the sutures a padded collar of X-ray film was placed around their necks. The control wound was left to heal under normal conditions before the second, the test wound, was inflicted, and this wound healed under experimentally altered conditions.

Index of the rate of repair. The tensile strength (T.S.) of the healing wound *in situ*, generally recognized as reflecting the rate of repair, was determined 5 days after wounding. A tensiometer, designed and manufactured at our Institute, was employed. Fig. 65 (facing p. 249) shows the design currently in use, an improvement of the prototype described by Sandblom, Petersen and Muren (1953). The device is based on the same principle as a thread-tester used in the textile industry. After the needles of the distender have been inserted at the edges of the wound, the claws of the distender are subjected to a pull until the wound is disrupted. This critical magnitude of pull is automatically indicated on the circular scale. The accuracy of determination of the tensile strength of healing wounds is $\pm$ 2·0 g. The T.S. is the force which just suffices to disrupt the wound, and the tensiometer measures T.S. in grammes of pull. For each determination the sutures were removed and the T.S. of three 1 cm lengths of the wound was determined. The three readings were averaged, and the mean was taken to represent the T.S. of the whole wound. The difference in T.S. of control and test wounds are expressed as a percentage of the control value.

HFC *of wound and granulation tissue.* Granulation tissue was allowed to grow into plastic sponges ('Ivalon'), $3 \times 1 \times 0\cdot3$ cm, placed between the skin and underlying muscle. Wound tissue, as sampled, included 2–3 mm of the edges of the wound, and comprised the wound tissue proper admixed with newly formed granulation tissue. The tissues were minced, incubated with ^{14}C-histidine and assayed for HFC in the usual way.

Granulation tissue and wound tissue were obtained at various intervals after wounding, and their HFC was compared with that of abdominal skin from the same rat (Fig. 66). In wound tissue

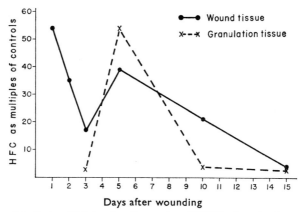

FIG. 66. Rate of histamine formation (HFC) of excised wound tissue and granulation tissue expressed as multiples of the activity in whole skin from the same rat. Each figure is the mean of determinations in two rats (Kahlson, Nilsson, Rosengren & Zederfeldt, 1960, *Lancet* ii, 230).

excised 24 hours after infliction of a wound the level of HFC was 50–60 times the level in control skin. The HFC of wound tissue declined on the 2nd and 3rd days, but was high again on the 5th day coincident with the rise in HFC of the granulation tissue. This was not measured during the first two days because at that time granulation tissue had not penetrated the plastic sponges. In the granulation tissue the HFC reached a peak at the 5th day and then fell steeply. The high activity noted in wound tissue at the 5th day is likely to reside mainly in the admixed granulation component. For comparison the HFC of various tissues of the rat (dog and cat included) is given in Table 68.

Table 68. HFC *in vitro*, expressed as amount of radioactive histamine in counts/min per gram rat tissue formed from ^{14}C-histidine

Foetal liver*	14,000–23,000		
Stomach†	370	–	980
Mast-cell tumour (dog)‡	470	–	530
Granulation tissue, 5th day	100	–	225
Wound tissue, 1st day	140	–	155
Lung†	90	–	150
Hypothalamus (cat)§	90	–	125
Total brain†	20	–	35
Skin of the back	2	–	10

(Figures quoted from: * Kahlson *et al.* (1960).
† Kahlson *et al.* (1958).
‡ Lindell *et al.* (1959).
§ White (1959).
(Kahlson, Nilsson, Rosengren & Zederfeldt, 1960,
Lancet, **ii,** 230.)

Histamine content of excised tissue. In four rats which were not given injections to alter the HFC, granulation tissue, wound tissue, and intact skin from the corresponding opposite region of the back, were examined for histamine at the 5th day of wounding. The HFC of the reparatively growing tissues was then high, as shown in Fig. 66. Nevertheless, the histamine content of these tissues was much lower than that of the intact skin (Table 69). This applies particularly to granulation tissue of which the histamine content is lower than any other tissue of the rat so far examined. The histamine formed so rapidly is not held by these tissues, indicative of nascent histamine. It will be recalled that the property of high HFC and concomitant low histamine content is characteristic also of embryonic tissues.

Table 69. Histamine content µg base/g in rat tissues on the 5th day after inflicting the wound (extracts assayed on guinea-pig ileum)

Rat No.	1	2	3	4
Wound tissue	5·4	8·7	6·6	6·5
Granulation tissue	1·7	1·6	1·4	1·1
Skin of the back	16·5	13·3	20·6	18·0

(Kahlson, Nilsson, Rosengren & Zederfeldt, 1960, *Lancet*, **ii**, 230.)

HFC *examined* in vivo. In these experiments four incisions of standard type were made simultaneously in the skin of the back. The daily excretion of free histamine was estimated by collecting the urine and assaying it directly on a piece of guinea-pig gut, as

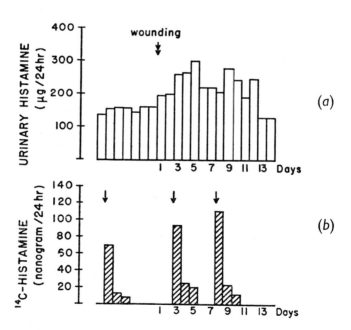

FIG. 67. (*a*) Daily urinary excretion of histamine in a female rat. (*b*) Arrows indicate subcutaneous injection in the same rat of radioactive histidine (110 µg per injection); cross-hatched columns represent urinary excretion of radioactive histamine resulting from these injections during the three subsequent days (Adapted from Kahlson, *et al.*, 1960; *Lancet*, **ii**, 230).

described in Chapter 5. As a further means of measuring the rate of histamine formation, ^{14}C-histidine was injected subcutaneously, and the excretion of ^{14}C-histamine was measured. Fig. 67 shows that the urinary histamine increases on the day of wounding, reaches a peak on the 5th day, and then returns to normal on the 13th day. The *in vivo* changes in the rate of histamine formation, as expressed by its urinary excretion, agree well with the corresponding observations *in vitro* (Fig. 66). In the experiment of Fig. 67 ^{14}C-histidine, 110 μg, was injected before and after wounding. Injection before wounding resulted in 700 counts/min ^{14}C-histamine excreted in 72 hours, and injections at the second and 7th days after wounding gave 1100 and 1160 counts/min, respectively. Similar results were obtained in three other rats. There was good agreement between results obtained with the simple non-isotopic method and the elaborate isotopic procedure.

Normal rate of repair (controls). Observations on ten normal rats (Fig. 68) show considerable variation in the rate of normal healing. This did not affect our study because in most animals identical wounds inflicted as described, and left untreated, healed with no apparent difference in rate. Moreover, healing both under normal and experimentally altered conditions was observed in the same individual. In the group of ten rats the mean T.S. value for test wounds inflicted after handling and after saline injections did not differ significantly from the mean for control wounds (Fig. 68).

Wound healing at elevated HFC. In the chapters on gastric secretion (Chapter 9) and anaphylaxis (Chapter 14), experiments are described in which experimental lowering of tissue histamine content was accompanied by an increased histidine decarboxylase activity. This phenomenon we tentatively explained as the result of a feed-back coupling between end-product and enzyme (end-product repression). This formal relationship, whatever its nature, can easily be demonstrated in the skin of the rat in which the level of histidine decarboxylase increases on injections of the histamine liberators 48/80 or polymyxin B (Schayer, Rothschild & Bizoni, 1959; Schayer & Ganley, 1959). Schayer and his colleagues did not specifically comment on the nature of the phenomenon, but we have confirmed the observations.

The HFC of rat skin is very low (Table 68). On draining the

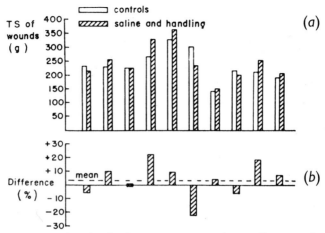

FIG. 68. (*a*) Each pair of columns represents the tensile strength of the control wound and of the test wound in a group of 10 rats (saline, 3 ml. intraperitoneally, was injected on 3 successive days before the test wound was inflicted); in (*b*) the effect of this treatment on the tensile strength is expressed as difference (%) against the control (Kahlson *et al.*, 1960, *Lancet* **ii**, 230).

skin of its histamine content by injecting 48/80 on three successive days, the HFC rose to 5–6 times normal. Observations on ten rats in which the test wound was inflicted 24 hours after the last 48/80 injection are presented in Fig. 69. Healing was significantly more rapid in the skin with an elevated HFC.

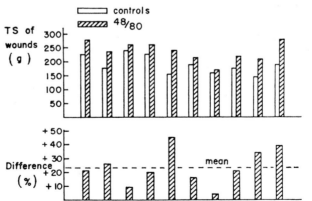

FIG. 69. Effect of injections of 48/80 (Kahlson *et al.*, 1960, *Lancet* **ii**, 230).

Polymyxin B, a more specific histamine liberator than 48/80, also evoked elevation of HFC. In rats under the influence of polymyxin B we found the rate of healing accelerated by a mean of 18%.

Repair in the phase of 'overshoot'. It has been described in Chapter 8 that on discontinuing measures which inhibit histamine formation a period of rebound (overshoot) ensues during which histamine formation is increased 2–4 times for a week or longer. Observations during this phase appear particularly significant because the tissues are cleared of drugs, i.e. the elevated HFC constituted the sole tissue change relevant in this connexion. Fig. 70 illustrates the singularly great increase in rate of healing in the period of elevated HFC which followed on discontinuing semi-carbazide and restoring pyridoxine to the diet. This easily reproducible experiment plainly revealed the association between HFC and anabolic processes.

Repair at lowered HFC. So far only one means has been found to inhibit histamine formation strongly (*ca.* 80%), over long periods

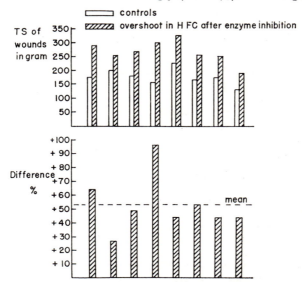

Fig. 70. Healing during the 'overshooting' phase (Kahlson, *et al.*, 1960, *Lancet* **ii**, 230).

in vivo, namely semicarbazide in conjunction with a pyridoxine deficient diet. This procedure inhibits decarboxylases besides histidine decarboxylase. Nevertheless, this procedure has brought forth useful information, as discussed in Chapter 8. And there is no evidence that decarboxylases other than histidine decarboxylase are involved in the healing of skin wounds. In the experiment illustrated in Fig. 71 tissue HFC was lowered to about 20% of normal. At this stage healing was greatly retarded. Inhibition by about 50%, obtained by simply feeding the rats on the pyridoxine deficient diet, did not significantly alter the rate of healing. It should be recalled that this moderate degree of inhibition did not retard growth of the rat foetus, whereas 80% inhibition did inhibit growth.

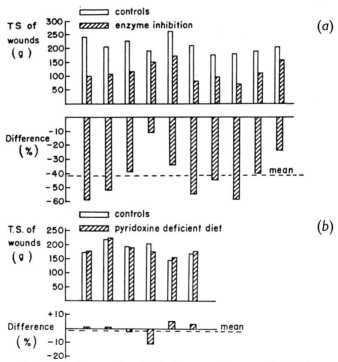

FIG. 71. (*a*) Healing under the influence of enzyme inhibition with semicarbazide superimposed on a pyridoxine-deficient diet; (*b*) the effect of the deficient diet alone (Kahlson *et al.*, 1960, *Lancet* **ii**, 230).

Ineffectiveness of extracellular histamine and anti-histamine drugs.
The histamine formed at high rates in granulation and wound
tissue will diffuse into the extracellular spaces, and thence enter
the blood stream to be distributed all over the body. As a result,
the granulation and wound tissues are exposed to extracellular
histamine in considerable concentration. It could be alleged that
histamine acting outside rather than inside the cells constitutes
the effective stimulus to reparative growth. This appears unlikely.
We flushed the healing wound with extracellular histamine derived
from 'long-acting histamine', an oily suspension of histamine
dipicrate deposited under the skin. With daily injections of this
material the histamine concentration in the blood was higher
than that resulting from four wounds, as evident from the histamine excretion which was 2–4 times larger than that in Fig. 67.
Exposure to extracellular histamine in this manner did not
significantly influence the rate of repair (Fig. 72).

During the last week of pregnancy the maternal tissues are
exposed to high concentrations of histamine produced by the
foetuses. The rate of healing of skin wounds exposed to histamine
from this source was not significantly different from that in the
non-pregnant state. Administration of anti-histamine drugs did
not alter the rate of healing (these two latter observations are by
Sandberg, unpublished).

In conclusion: The mode of action of nascent histamine cannot
be fulfilled by extracellular histamine, nor is its action antagonized
by anti-histamines.

HFC *and rate of collagen formation.* It is known that the tensile
strength of skin wounds is related to its collagen content. The rate
of formation of collagen can be determined by implanting polyvinyl sponges in the skin wound and estimating the amount of
hydroxyproline (an obligatory constituent of collagen) contained
in the sponge. Our colleague Sandberg (1964a) determined the
rate of formation of collagen (hydroxyproline) in skin wounds in
rats before and after elevation of skin HFC following injections of
48/80 or polymyxine B. Fig. 73 shows that collagen is formed at
higher rates, particularly at days 3 to 11 of wounding, at the state of
elevated HFC following administration of 48/80. HFC elevation
following polymyxine B similarly enhanced collagen formation.

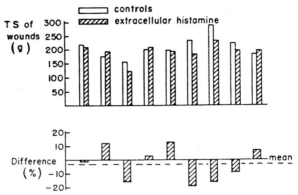

FIG. 72. Healing under the influence of extracellular histamine (Kahlson *et al.*, 1960, *Lancet* **ii**, 230).

The effect of cortisone on collagen formation and wound tissue HFC. It is well known that the administration of cortisone in large doses inhibits the rate of wound healing (for references see Dempster,

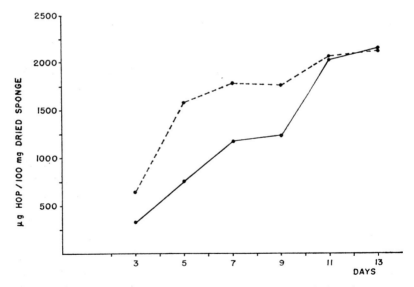

FIG. 73. Total amount of hydroxyproline in granulation tissue in rats. Control sponges ————, experimental sponges — — — (Sandberg, 1964*a*, *Acta Chir Scand.* **127**, 9).

1957). Schayer and his colleagues have shown that cortisone inhibits histamine formation in the rat skin (Schayer, Smiley & Davis, 1954). Sandberg (1964*b*) found that under the influence of cortisone the concentration and the amounts of hydroxyproline in the granulation tissue growing into implanted polyvinyl sponges was greatly reduced (Fig. 74). Further, Sandberg and Steinhardt (1964) demonstrated that administration of cortisone restrained the elevation of HFC which normally is engendered in the tissues of the wound; from this result they concluded that cortisone retards healing by way of its restraining effect on HFC.

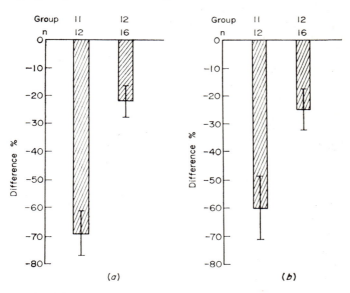

FIG. 74. Mean percentage differences ± s.e.m. of (*a*) hydroxyproline concentration (μg/100 mg dried tissue) and (*b*) total amount of hydroxyproline (μg/100 mg dried sponge) in control and experimental sponges. Cortisone treatment before implantation of experimental sponges, groups 11 and 12 at days 5 and 7, respectively (Sandberg, 1964*b*, *Acta Chir Scand.* **127**, 446).

HFC *in human wound tissue.* Normal human skin forms histamine, as first demonstrated by Demis and Brown (1961). Thanks to the co-operation of Dr. Gösta Jönsson, Professor of Surgery, University of Lund, we were given the opportunity to investigate the HFC of human wound tissue (Kahlson, Rosengren & Steinhardt,

1963*b*). In these experiments a 2–3 cm long incision was made through the entire thickness of the skin of the lower abdominal region and closed by sutures under local anaesthesia. Circular pieces of skin, including the full length of the incision, were excised at 24 or 48 hr after wounding. Normal tissue to serve as controls was obtained from excised pieces at sites maximally distant from the incision. The HFC was determined isotopically in the usual way. In the majority of eight patients investigated the HFC of wound tissues was considerably higher than in the specimens of intact skin.

A similar study was carried out by Lindell, Westling and Zederfeldt (1964) in nine patients operated on for bilateral varicose veins in a 2-stage procedure under general anaesthesia. The HFC (μg ^{14}C-histamine formed in 3 hr per gram) was determined in normal skin (skin I, Table 70), in normal skin after the first operation stage (skin II) and in excised skin wound tissue. In the wound tissue there was a tendency for HFC to reach a peak on the second and third day after the incision. This result in man agrees fairly well with the pattern of changes in rat wound tissues shown in Fig. 66.

Table 70

Age of patient	Sex	Wound age (days)	Histamine-forming capacity		
			Skin I	Skin II	Wound
53	♂	1	11·3	0	0
55	♀	1	0	5·6	27·5
39	♀	2	0	0	30·3
56	♂	2	30·4	15·3	65·8
67	♀	3	0	0	73·5
39	♂	3	0	0	48·5
61	♂	4	0	4·8	42·3
39	♂	4	0	0	15·3
69	♂	11	1·6	0	14·6

(Lindell *et al.*, 1964, *Proc. Soc. exp. Biol. Med.* **116**, 1054.)

Human skin mast cells form histamine (Lindell, Rorsman & Westling, 1961*a*) and for this reason it appeared essential in our study in 1963 to investigate the possibility of changes in the number of mast cells in the skin specimens under study. The determinations summarized in Table 71 were done by Dr. H.

Rorsman of the Dermatology Clinic, Lund, and they show that the number of mast cells in the immediate region of the wound is not higher than at the control sites. In this connexion it should be mentioned that in the rat the number of mast cells is reported to be decreased in the region of skin wounds in the first 24 hr (Wickmann, 1955).

Table 71. Number of mast cells per unit in tissues of the wound and control sites

Age	Sex	Patient Diagnosis	Wound A	B	C	Control A	B	C
45	♂	Fistula urethrae	238	713	262	152	262	279
69	♂	Hyperplas. prostatae	457	812	172	421	795	426
69	♂	Hyperplas. prostatae	437	1000	156	600	746	254
62	♀	Cancer ves. urin.	388	1279	271	354	853	213
		Mean:	380	951	215	382	664	293

A: Section through total cutis.
B: Section through uppermost dermis.
C: Section through lower dermis.
(Kahlson *et al.*, 1963*b*, *Experientia*, **19**, 243.)

Only a little information is on record on the problem of the rate of histamine formation in mast cells. Particularly scanty is the information about human mast cells. In the investigation by Lindell, Rorsman & Westling (1961*a*) on histamine formation in urticaria pigmentosa the HFC of normal skin from the abdomen and the back was so low that the amounts of ^{14}C-histamine determinable *in vitro* were not formed from ^{14}C-histidine. Even in the skin where the mast cell infiltration was dense, the HFC was not high compared with the high values observed in other tissues (Table 72). In this table from the report by Lindell *et al.* (1961*a*) 1 μg of ^{14}C-histamine formed from ^{14}C-histidine gave 600 counts/min and the method used was in the main the one employed in our laboratory. In mice Kahlson, Rosengren and Steinhardt (1963*a*) found HFC in the order of 25 counts/min in 10^5 peritoneal mast cells. Rat peritoneal mast cells are reported to be rich in histidine decarboxylase (Schayer, 1956*a*).

Tissue HFC *in cutaneous burns.* Burning the skin rapidly induces

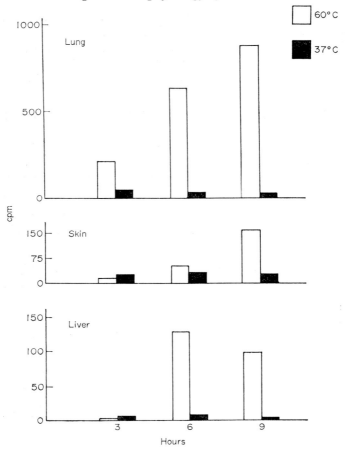

FIG. 75. Histamine formation in lung, skin and liver tissues in mice after cutaneous burns. The activity of histidine decarboxylase is given as count/min of [^{14}C]-histamine formed from [^{14}C]-histidine (ordinate). Abscissa: hours after immersion in water at 60° and 37° respectively (Johansson & Wetterqvist, 1963, *Med. exp.* **8,** 251).

reparative processes, i.e. healing. From the experiments cited above it would *a priori* appear that burned skin should generate elevated histidine decarboxylase activity. Relevant experiments have been done by Schayer and Ganley (1959). They immersed anaesthetized mice to the shoulder in water at 60 °C for 25 seconds, and the controls were dipped in water at 37 °C. The skin, excised six hours after burning, displayed a HFC 3-times the control value.

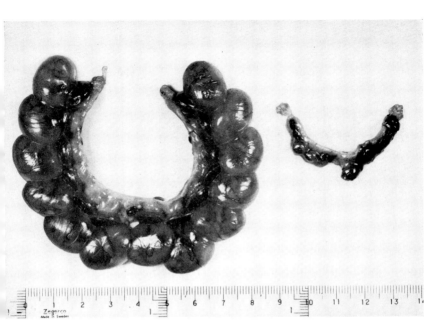

FIG. 61. The uterine contents of two rats at the 19th day of pregnancy. *Left*, undisturbed pregnancy; *right*, after enzyme inhibition (Kahlson & Rosengren, 1959, *Nature, Lond.* **184,** 1238).

FIG. 65

[*facing p. 249*

Table 72

Case	Diagnosis	Mast cells/mm³		Histamine content (μg/g)		Histamine formation (counts/min) (weight)	
		Normal skin	Pathologic skin	Normal skin	Pathologic skin	Normal skin	Pathologic skin
I	Urticaria pigmentosa	12,000	110,000	—	126	0 (0·13 g)	39 (0·13 g)
II	Urticaria pigmentosa	17,000	320,000	—	360	0 (0·33 g)	230 (0·30 g)
III	Leukaemia	—	4000	9	10	0 (0·20 g)	0 (0·20 g)

(Lindell *et al.* 1961a, *Acta Dermat. Venerolog.* **41**, 277.)

Similar and somewhat extended experiments have been carried out by Johansson and Wetterqvist (1963). The procedures were the same as in the study by Schayer and Ganley, except that tissues were excised at 3, 6 and 9 hours after burning, and the HFC of lung and liver was also determined. The results are shown in Fig. 75. The elevation of HFC in liver and lung on burning the skin appears puzzling and recalls the phenomenon of elevated tissue HFC in moribund mice which is described under a separate heading (Chapter 15).

Clinical implications. In theory it would appear that there is a possibility that elevation of skin HFC by means such as polymyxin B might be helpful in accelerating healing and the formation of wound tissues in homologous skin grafting and tissue repair after burns.

Keele and Armstrong (1964) suggest that histamine formed in granulation and wound tissue may contribute to the sensation of itching often felt during the healing of wounds of the skin, and they have frequently recorded this phenomenon during the healing of a blister base on about the 3rd-5th day, which corresponds to the days when histamine formation is high in rat and human skin wounds.

I

20

HISTAMINE FORMATION IN BLOOD CELLS AND BONE MARROW

THE bone marrow contains a heterogeneous population of cells which, however, possess metabolic features common to all types of marrow cells: high rates of cell growth and protein synthesis (reviewed by Lajtha, 1957). Our study of the bone marrow was preceded by an investigation of histamine formation in human blood by our colleagues Lindell, Rorsman and Westling (1961b) employing Schayer's isotopic technique. Like Hartman, Clark and Cyr (1961) they found histamine formation in blood samples from patients with myelogenous leukaemia. Lindell *et al.* showed that the activity was present in the buffy coat and there was a significant correlation between the number of mature basophil cells and the rate of histamine formation. They emphasize the characteristic common to tissue mast cells and basophil leucocytes, namely, that both cell types can synthesize their histamine. No histamine formation was obtained on incubation of plasma.

Red bone marrow from the humerus, femur and tibia of rats was investigated for histamine formation by Kahlson and Rosengren (1963). The object was to assess approximately the proportion of whole-body histamine formation that takes place in the bone marrow, cytological examination of the tissue was thus not considered essential. Results obtained in eleven experiments are summarized in Table 73; in each of the three last experiments of the table, tissue from two rats was pooled. These results are noteworthy in that they show that the HFC per gram of bone marrow exceeds even that of the foetal rat liver and is about 10 times higher than in the gastric mucosa. It is thus obvious that a very large proportion of histamine formation in the rat takes place in the bone marrow. This result made us alter our hitherto misguided belief that in the rat about half of the whole-body HFC resides in the stomach, a topic which is discussed in Chapter 9.

Table 73. Histidine decarboxylase in rat bone marrow

Age in days	Sex	Amount of tissue incubated (g)	(counts/min per gram)
41	♂	0·07	30,000
43	♀	0·11	20,000
45	♀	0·08	26,000
46	♀	0·10	21,000
48	♀	0·11	40,000
52	♂	0·17	21,000
54	♂	0·13	24,000
63	♀	0·06	58,000
76	♂	0·16	26,000
140	♀	0·18	25,000
140	♀	0·23	31,000

(Kahlson & Rosengren, 1963, *Experientia*, **19**, 182.)

Håkanson (1964) has confirmed the presence of histidine decarboxylase in the bone marrow of the rat and that the enzyme, in some investigated respects, behaves as foetal histidine decarboxylase.

In the guinea-pig, histamine formation in the blood and bone marrow was investigated by Aures, Winqvist and Hansson (1965) by measuring the $^{14}CO_2$ liberated from DL-histidine-1-$^{14}COOH$. They found that the histamine levels and the histidine decarboxylating activity correlated to the number of basophils. Treatment of guinea-pigs with foreign protein resulted in a 10-fold increase of both histamine concentration and histidine decarboxylating activity in the blood and bone marrow.

21

HISTAMINE FORMATION IN MALIGNANT AND REGENERATIVE GROWTH

AFTER the eminently high HFC had been recognized as part of certain kinds of rapid tissue growth, it appeared likely that a similar phenomenon might be found in malignant growth also. In fact, this is so, although some rapidly growing tissues seem to form the amines putrescine and spermidine in the place of histamine.

Mastocytoma in dogs and mice. The mast cells are very rich in histamine, but their HFC is not particularly high. In our laboratory a typical mastocytoma was found in the abdominal skin of a dog in 1958. The tumour was investigated by Lindell, Rorsman and Westling (1959) who found the HFC to be high, although lower than in rat gastric mucosa and foetuses. A mouse mastocytoma also displayed high histidine decarboxylase activity (Hagen, Weiner, Ono & Lee, 1960). In a study by Werle, Schauer and Bühler (1963), mast cells of normal mouse tissues were found to be rich in histamine, but their histidine decarboxylase activity was very low. In Furth mastocytoma, by contrast, the histidine decarboxylase activity was high and the histamine content of individual mast cells was low. It is set out later that in tumours with high HFC the histamine content is low.

Rat hepatoma. This transplantable tumour grows rapidly in rats of the August strain. Mackay, Marshall and Riley (1960) noted in female rats bearing a hepatoma implanted subcutaneously, that the urinary histamine rose steeply from a normal level of 30–40 µg per rat in 24 hr to values in excess of 1 mg per rat per day. On removal of the tumour, the histamine excretion at once fell to the normal level. The Scottish investigators implanted portions of the tumour into the controls whose urinary histamine thereupon rose. It will be described later that the rat liver, under certain circumstances, easily generates high histidine decarboxylase activity.

Rat mammary carcinosarcoma. In female rats bearing subcutaneous transplants of the Walker 256 carcinoma, Håkanson (1961), employing Schayer's isotopic method, found HFC values in the range 94 to 154 ng ^{14}C-histamine formed per gram tissue in 3 hr. For comparison, in normal rat mammary tissue, excised from pregnant rats 20–21 days after mating, the HFC was too low to permit exact determinations. Mast cells were not found in the tumour tissue investigated, and its histamine content was very low. Hallenbeck and Code (1962), applying Waton's non-isotopic method to experiments on this tumour, found that histamine formation was barely detectable by the method, and, more significantly, they found no consistent change in the excretion of histamine when the tumours were implanted or growing rapidly, or after their removal and subsequent recurrence. Further results with a rat mammary tumour will be described in the following section.

Ehrlich ascites tumour in mice

The changes in HFC occurring in this tumour have been more thoroughly studied than in any other rapidly growing tissue. The Landschütz I tumour growing in the mouse peritoneal cavity as single free cells, sometimes referred to as the Ehrlich ascites tumour, lends itself particularly well for study because in this tumour it is possible to determine (1) the total number of cells present; (2) the actual phase of growth expressed as the frequency of cells in mitosis at different stages of the experiment; (3) the HFC, the histamine content and the binding capacity of these particular cells exclusively.

In this study by Kahlson, Rosengren and Steinhardt (1963*a*) a hyperdiploid Ehrlich ascites tumour line, also known by the synonyms Landschütz I tumour (Tjio & Levan, 1954) or ELD (Hauschka, Grinnell, Révész & Klein, 1957) was investigated. For details of procedures our report should be consulted.

HFC *and cell multiplication.* A typical observation is illustrated in Fig. 76, which shows that the rate of cell multiplication in the tumour is high during the first 3 days after inoculation, as indicated by the mitotic index and the growth curve. In later stages the rate of cell multiplication regresses steadily, whereas the total tumour bulk increases continually throughout the experiments, in accord with previous reports (Hauschka *et al.*, 1957). It will be seen from the figure that during the phase of rapid cell multiplication the histidine decarboxylase activity per unit mass of tumour is very high. In later stages, with decreasing mitotic index, the HFC falls correspondingly and is nearly zero when cell multiplication is in abeyance. At this stage the individual cell merely increases in bulk, a kind of growth which in this particular tumour tissue is not conducive to generating histidine decarboxylase.

Participation of mast cells in HFC *elevation.* Since the ascites fluid rinsings contained mast cells besides tumour cells, and the peritoneal mast cells contain histidine decarboxylase (Schayer, 1956*a*), it was essential to investigate the part played by the mast cells in the combined fractions of such cells and tumour cells. The HFC of peritoneal mast cells was in the order of 25 counts/min in 10^5 cells. On the first days of tumour growth, on using a relatively small inoculation dose (10^6 tumour cells), about half the HFC of peritoneal rinsings resided in mast cells. With a higher inoculation dose (3×10^6) the contribution of mast cells to the HFC was only about 25% and there is a tendency for the number of mast cells to decrease as the tumour grows. These observations do not detract from the conclusion that the elevation of HFC noted in the pooled cell population occurs in the tumour cells, especially in the early phase of growth in which the mitotic index is high.

Characteristics of the tumour enzyme. Characteristics of the rat foetal histidine decarboxylase have been described in Chapter 6. The object here was to see to what extent the decarboxylase of the tumour cells resembles the enzyme contained in embryonic tissue. Tumour cells were obtained on the third day after inoculation (inoculation dose: 4–5×10^6 tumour cells) when the HFC almost exclusively depended on the tumour cells. With the isotopic method and with the enzyme: substrate ratio usually employed,

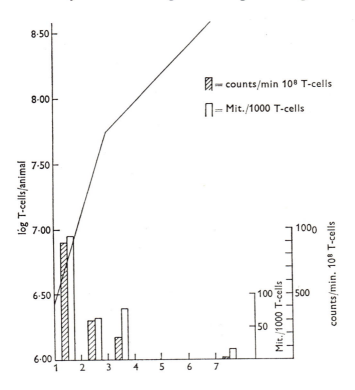

Fig. 76. The ordinates represent, respectively, number of cells in the tumour (T-cells), number of mitoses per 1000 tumour cells, histidine decarboxylase activity expressed as ^{14}C-histamine (counts/min) formed on incubating the cells for 3 hr with ^{14}C-histidine. The abscissa denotes days after inoculation of the tumour (Adapted from Kahlson *et al.*, 1963a, *J. Physiol.* **169**, 487).

the pH optimum of the tumour enzyme was in the range 7·0–7·4. Benzene, which activates DOPA-decarboxylase, did not increase the activity of the tumour enzyme. The two inhibitors which we use as a means of enzyme identification, α-methylhistidine and α-methyl-DOPA in the concentration 10^{-3} M, inhibited the rate of histamine formation by 90% and 7%, respectively. Destruction of the cell structure by freezing for 24 hr reduced the enzyme activity to 40% of the activity as determined in whole cells.

Finally, the histamine content of the tumour cells was low and their capacity to bind the histamine formed was less than in mast cells. In all three respects the tumour enzyme was found to be identical with foetal histidine decarboxylase.

HFC *in various tissues of moribund mice.* Observations by Schayer (1962*b*) on moribund mice are mentioned in Chapter 15. From these observations he suggested that the increase in enzyme activity 'was due to systemic effects as well as to local irritant effects of bacteria'. We therefore investigated whether a similar elevation in enzyme activity would occur in mice dying from the Ehrlich ascites tumour. We determined the HFC of various tissues in moribund tumour-inoculated mice at different intervals after inoculation. The HFC was substantially elevated in the tissues of inoculated mice, except in the kidney (Table 55). The means by which the nearly universal elevation of tissue HFC is brought about in the moribund mouse appears obscure and presents a subject for further study. The complexity of the phenomenon may be exemplified by the observation that in the pregnant mouse in which the foetuses produce high HFC, the kidney is the sole organ in which HFC elevation occurs (Rosengren, 1963), but in the moribund kidney it is the exact opposite.

HFC *in solid tumours of inoculated mice.* Among the great number of mice inoculated with the Ehrlich ascites tumour we found solid tumours in three moribund mice. Tumours I and III consisted of infiltrations of the abdominal wall and tumour II of infiltration of the omentum. The tumours were removed, their HFC was determined and the total number of mitotic figures was counted. The observed values for HFC are given as counts/min per gram tissue and the mitotic figures in brackets: tumour I, 13,000 (30); tumour II, 1200 (17); tumour III, 1600 (14). It thus appears that the HFC was highest in the tumour with the highest number of mitotic figures.

Histamine excretion. In the pregnant mouse, in which the foetuses produce very large amounts of histamine, increased histamine excretion can easily be demonstrated. In mice bearing the ascites tumour the urinary histamine was not detectably increased. Although the HFC of the tumour cells was high, the total tumour mass is small, and its histamine formation is likely to represent only a very small fraction of the whole-body histamine formation.

Attempts at inhibiting tumour growth. In Chapter 8 it has been described that in the rat histamine formation can be strongly inhibited *in vivo* by α-methylhistidine or by semicarbazide in combination with a pyridoxine-deficient diet, and moderately inhibited by pyridoxine deficiency alone. As supplies of α-methyl-histidine in amounts required for strong inhibition *in vivo* in tumour inoculated mice were no longer available, and semicarbazide proved highly toxic in mice, we resorted to the possibility to inhibit moderately the HFC by feeding a diet deficient in pyridoxine. Groups of mice were subjected to the pyridoxine-deficient diet for 4, 6 and 8 weeks before inoculation, and this diet was maintained during the course of the actual experiment. The number of tumour cells, the mitotic index and HFC were determined and compared with a group of controls fed adequately and inoculated at the same time with the same dose. The results are shown in Fig. 77. In group II a correlation between retardation of growth and the fall in mitotic index during the first 3 days of inoculation is particularly noticeable. The HFC of the tumour material obtained from pyridoxine-deficient mice on any of the first 3 days after inoculation, determined *in vitro*, was reduced to

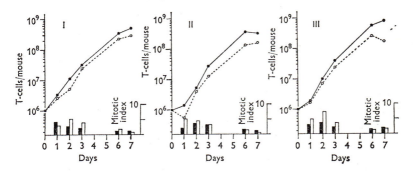

Fig. 77. Growth curves (log. scale) and mitotic index of ascites tumour in groups of mice fed adequately ☐ (———) and fed on a pyridoxine-deficient diet ■ (— — — —) for 4 weeks (I), 6 weeks (II) and 8 weeks (III) before inoculation. The abscissa denotes days after inoculation (Kahlson *et al.*, 1963a, *J. Physiol.* **169**, 487).

20–30% of the control tumour material. It should be recalled that in order to obtain significant inhibition of growth in the rat the inhibition of tissue HFC must exceed 50%.

HFC of human nasal mucosa

This investigation is interesting since it is one of the few which has so far been done on human hypertrophically growing tissue. Polypus tissue was excised and its HFC was determined isotopically *in vitro*. In seven specimens the HFC values were 44, 112, 113, 182, 237, 260 and 360 counts/min per gram tissue. In normal nasal mucosa, which was the nearest control tissue, the HFC was below the level for accurate determination (Johansson, Lindell & Westling, 1962, unpublished).

Pancreatic tumour in Zollinger–Ellison's syndrome

A specimen of this kind of tumour, excised in a previously healthy woman, aged 31, was investigated by Dotevall, Lindell and Westling (1963). An extract of the tumour showed gastrin-activity, ascertained by determination by Professor Gregory, Liverpool. The HFC in the tumour, determined *in vitro*, was slightly higher than in the surrounding parts of the pancreas. However, there was indication of a high rate of histamine formation *in vivo* in that the urinary excretion of histamine was abnormally high and the increased histamine excretion fell to the normal low value on the removal of the tumour (Fig. 78). The fall in histamine excretion is presumably due in part to elimination of histamine formed by the tumour and in part to restrained gastric mucosal histamine formation following reduced stimulation by gastrin produced by the tumour. The authors suggest that in patients suspected of the Zollinger–Ellison syndrome, the urinary excretion of histamine should be examined. In passing, on the assumption that this interpretation be correct, these observations would show, for the first time in man, that endogenous gastrin was capable of accelerating gastric mucosal histamine formation in such high degree that part of the resulting histamine appeared in the urine.

HFC in virus-induced rat sarcoma

The tumour was induced in 1962 by Rous chicken sarcoma virus strain S.R. and investigated by Ahlström, Johnston and Kahlson

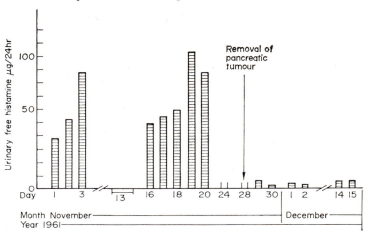

Fig. 78. Urinary free histamine in µg base per 24 hr and gastric acid secretion in m-equiv/l. before and after operation. Upper range of normal urinary free histamine in women is 30 µg/24 hr (Dotevall *et al.*, 1963, *Acta med. scand.* **174**, 325).

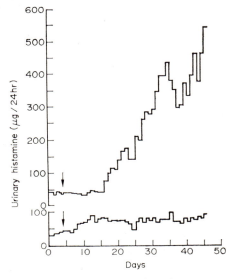

Fig. 79. Top curve illustrates daily urinary excretion of free histamine (in terms of the base), in a female tumour-bearing rat, inoculated at the arrow. Bottom curve shows results from an inoculated female litter-mate in which no tumour developed. (Ahlström *et al.*, 1966, *Life Sci.* Oxford, **5**, 1633).

(1966) as follows: the animals used were young commercial albino rats which had been kept for years as a closed colony at the Pathology Institute, University of Lund. For transplantation, a piece of tumour was suspended and injected into a thigh of rats about 4 weeks old. Growth was palpable after 4–7 days.

Histamine formation in vivo. The excretion of histamine was followed in tumour-bearing rats and in control animals in which tumours failed to develop. Between 7 and 12 days after inoculation the amount of histamine in the urine of the former group began to increase and continued to do so steadily as the tumour grew. The histamine output of the controls rose only slightly, reflecting normal body growth. The results of a typical experiment are presented in Fig. 79.

Histamine formation in vitro. The HFC was determined isotopically in twenty-two tumour-bearing rats and the results are summarized in Fig. 80. There is a tendency for younger tumours to display the highest HFC, but there is no strict correlation between tumour age and HFC. Several factors may contribute to the imperfect correlation. In the particular instance of the oldest tumour investigated, widely divergent figures, 166 and 500 ng/g, were obtained for the HFC of two parts of the same tumour. The higher value was found in tissue taken from a nodule which appeared to be a new outgrowth from the main tumour. The HFC of this virus-induced sarcoma was high in comparison with most normal tissues. There is no normal tissue structurally related to this particular neoplastic tissue. In many of the tumours investigated the HFC exceeded the values found in wound tissue and granulation tissue. HFC values of some normal rat tissues are given in Table 7.

Mitotic index. The mitotic index, expressed as mitoses per 10^3 cells, was determined in seven tumours, on 4 μ thick paraffin sections, 5000 cells being counted in each tumour. The figures were 15·8, 13·8, 10·0, 8·0, 6·8, 4·6, 3·4, and the corresponding HFC values were 1340, 437, 458, 90, 115, 56, 312, expressed as ng/g. From this and from Fig. 80 it appears that there is a tendency for higher HFC to be associated with more frequent mitosis.

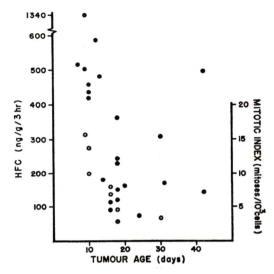

Fig. 80. Variation of HFC (filled circles) and mitotic index (open circles) with tumour age. Each circle represents a single determination in one tumour except for the 42-day-old tumour (Ahlström, *et al.*, 1966, *Life Sci.* Oxford 5, 1633).

Mast cells. The occurrence of mast cells was investigated in several specimens of tumour tissue, fixed in 4% lead acetate. No mast cells, or occasionally a single mast cell were seen inside the tumour tissue proper the HFC of which was determined. A sparse infiltration of mast cells was observed surrounding the tumour.

HFC *of the liver of tumour-bearing hosts.* At autopsy it was found that the liver was enlarged in rats bearing the virus-induced sarcoma. This phenomenon has been observed by investigators studying other types of tumour (for references see Ghadially & Parry, 1965). In rats bearing relatively old tumours, we found values of about four times the corresponding control figures, calculated as gram liver/g carcass. Obviously, here was another instance for examining whether this conspicuous enlargement of the liver was an example of growth accompanied by elevated HFC. Determinations were performed *in vitro*, isotopically as for the neoplastic tissue. The HFC of livers at various stages of tumour development was investigated. Rats from the same litter were taken as controls. Some of the controls were injected with dead tumour material, and in other controls (marked with a cross in Table 74) tumours had failed to develop. The table shows that the HFC of the enlarged liver was several times the corresponding control, ranging from 4 times in rats bearing young tumours to 14 times in the liver of a rat with a 30-day-old tumour. In one case there was no increase in HFC. A correlation appears to exist between the increase in liver weight and the magnitude of its HFC. The mitotic index of the enlarged liver was not determined in this study, but reports by other investigators present evidence of increased mitotic rate in the liver of rats and mice bearing various types of tumour (Annau, Manginelli & Roth, 1951; Hemingway, 1965).

Table 74. HFC of tumour and liver expressed as ng histamine formed/g wet weight

Tumour	Tumour age (days)	Tumour weight (g)	Tumour HFC (ng/g)	Ratio L/C*	Liver HFC (ng/g)
+	7	0·5	518	0·054	58
−				0·046	15
+	9	1·4	506	0·057	43
−				0·052	56
+	10	3·0	420	0·043	82
† −				0·034	17
† −				0·035	8
+	12	3·8	587	0·036	62
−				0·033	37
+	13	3·0	483	0·043	324
−				0·034	23
+	14	10·2	183	0·078	290
−				0·049	42
+	20	15·5	163	0·078	122
−				0·054	32
+	30	80·0	312	0·181	486
† −				0·037	32
† −				0·049	35
+	42	143·0	166;500‡	0·168	390
† −				0·042	43

* L/C = liver weight/carcass weight (weight of rat exsanguinated and minus tumour).
+ = Tumour-bearing.
− = Control.
† Tumour not developed.
‡ Samples from different parts of the same tumour.
(Ahlström *et al.*, 1966, *Life Sci.*, Oxford, **5**, 1633.)

Nature of the enzyme in the tumour and enlarged liver. The forma-tion of ¹⁴C-histamine from ¹⁴C-histidine *in vitro* by tissue of the tumour and the enlarged liver was strongly inhibited by α-methylhistidine but weakly by α-methyl-DOPA in the concentra-tion 10^{-3} M, 70–80% and 5–20% respectively. In the Ehrlich ascites tumour inhibition by the two compounds was respectively 90% and 7%. These results show that the enzyme mainly con-cerned in histamine formation in these three types of tissue is the same, namely histidine decarboxylase.

Histamine content. In the normal and in the enlarged liver with high HFC, the histamine content was very low, 1–5 μg(base)/g tissue. In the tumour tissue higher values for histamine content were obtained (up to 15 μg/g tissue), and in the tumour there was no correlation between content and HFC. It will be recalled that high HFC and low histamine content is characteristic of other rapidly growing tissues, e.g. foetal tissues and the tissues of healing skin wounds, tissues which generate nascent histamine.

The significance of high HFC *in the tumour.* In the study of virus induced rat sarcoma the tumour was transplanted into several successive individuals. The HFC of the tumour persisted at high levels throughout the line of transplants. It has been shown that generally after several transplant generations only those enzymic systems essential to energy production, growth and reproduction are retained in the tumour (Greenstein, 1954). Clearly, histidine decarboxylase ranks among this class of essential enzymes. The association between HFC and protein synthesis in rapidly growing normal and malignant tissues will be discussed in Chapter 22.

Walker carcinosarcoma, Chester Beatty Institute strain. On a visit to the Chester Beatty Institute in London, two rats bearing the Walker mammary carcinosarcoma of the strain of that Institute were obtained by the courtesy of Professor A. Haddow. The tumour was transplanted into adult female rats of a strain bred at the Institute of Physiology, Lund, and the tumour was investig-ated here by Marian Johnston (1967) on similar lines to those described for the virus-induced tumour. Johnston's results will be set out in some detail, because they differ from those obtained by Hallenbeck and Code (1962) and by Håkanson (1961) with this tumour in strains different from the Chester Beatty one.

Excretion of free histamine. Histamine excretion rose as the tumour grew in both Institute-strain and Sprague–Dawley rats. Increases of 3–8 times the pre-implantation values were observed and two examples from individual rats are shown in Fig. 81.

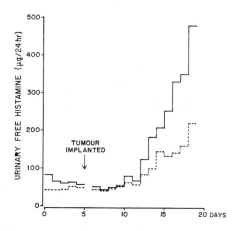

FIG. 81. Urinary histamine excretion patterns in two female rats implanted at the arrow with the Walker carcinosarcoma (Johnston 1967, *Experientia* **23**, 152).

Excretion of [14]*C-histamine after injection of* [14]*C-histidine.* Rats were injected subcutaneously with 50 μg ^{14}C-histidine before and after tumour implantation, and aliquots of the urines were assayed for ^{14}C-histamine for the three consecutive days after injection. The excretion of ^{14}C-histamine was higher during the course of tumour growth than in the control period before implantation (Table 75).

HFC *of tumour tissue.* From the observations *in vivo* it is evident that the carcinosarcoma formed histamine at considerable rate, and this was confirmed by determining HFC *in vitro.* The tumour tissue forms histamine at high rates (Table 76) which among normal rat tissues is exceeded only by the HFC of the gastric mucosa and foetal tissues (Table 7).

Table 75. Excretion of ^{14}C-histamine after injection of ^{14}C-histidine (Counts/min per total daily urine volume)

		Before implantation	12–15 days after implantation
Sprague–Dawley rat	1st day	195	240
	2nd day	28	115
	3rd day	14	78
Institute strain rat	1st day	83	259
	2nd day	18	73
	3rd day	6	81

(Johnston, 1967, *Experientia*, **23**, 152.)

HFC *of the enlarged liver.* In rats bearing this tumour, the liver was enlarged in a similar way to that described for the virus-induced tumours, and the HFC was found to be very much higher than in non-tumour-bearing controls. The HFC values and the liver weight are included in Table 76.

Table 76. Rate of histamine formation in tumour and liver tissue

Tumour age (days)	Tumour weight (g)	Tumour HFC (ng/g)	L/C*	Liver HFC (ng/g)
14	40	278	0·074	290
14	35	416	0·054	437
14	34	821	0·071	582
14	32	884	0·063	296
14	38	1090	0·073	1200
17	60	770	0·081	613
17	49	1190	0·064	371
17	37	1890	0·065	886
25	69	900	0·043	219
Controls			0·038	31
			0·044	27
			0·039	22
			0·049	25
			0·032	18

* L/C is the ratio liver weight/carcass weight (weight of rat exsanguinated and minus the tumour.)

(Johnston, 1967, *Experientia*, **23**, 152.)

Nature of the enzyme. As with other rapidly growing normal and malignant tissues, the effect of α-methylhistidine and α-methyl-DOPA on the rate of histamine formation in tumour tissue and liver of the tumour-bearing host were investigated in the study by Johnston (1967). The former compound produced 70–90% inhibition of enzyme activity in both tissues, but the latter had practically no inhibitory effect, thus indicating that the enzyme involved was histidine decarboxylase (Table 77).

Table 77. Effect of α-methylhistidine and α-methyl-DOPA (10^{-3} M) on the rate of histamine formation of tumour tissue and liver of the tumour-bearing host (ng histamine formed/g tissue)

	Tumour			Liver		
Rat	No addition	α-methyl-histidine	α-methyl-DOPA	No addition	α-methyl-histidine	α-methyl-DOPA
1	166	42	191	62	15	58
2	900	211	806	219	26	219
3	1550	381	1560	2530	354	2310
4	416	79	437	437	70	418

(Johnston, 1967, *Experientia*, **23**, 152.)

Parabiosis experiments

It was discovered in our laboratory that in the pregnant mouse the kidney of the mother is endowed with a singularly high HFC (Rosengren, 1963). This phenomenon could be understood on the assumption that the factor which brings about the generation of high HFC in the mouse foetuses is carried by the blood stream to the kidney which in the pregnant mother is highly susceptible to a 'trigger' agent. A device of this kind could, perhaps, operate in generating the high HFC in the kidney of the mother. Alternatively, the elevation of HFC may merely be an obligatory event in the process of rapid liver growth. It will be explained later that liver tissue has been found to grow rapidly without elevation of its HFC.

With a view to obtaining information on this problem, Johnston (1967) prepared pairs of parabiotic rats, using female litter-mates aged 5–6 weeks. This procedure is known to result in a degree of common blood flow. About 6 weeks after the operation, one partner from each of two pairs was implanted with the tumour. After a further 2 weeks both pairs were killed and the HFC of their tissues determined. A control pair which had not been implanted was also investigated. The HFC was elevated in the liver of the non-tumour-bearing partner also and here the value was intermediate between those found in the controls and in the tumour-bearing partner. This is consistent with the involvement of a blood-borne agent(s) in the mechanism causing elevation of HFC. In this case the effect must be greater in the tumour-bearing partner since the concentration of the agent in the blood of the counterpart will not be so high because of dilution, destruction, excretion and other conceivable circumstances. The experiments by Johnston have been repeated by Ishikawa, Toki, Moriyama, Matsuoka, Aikawa and Suda (1970) who, instead of parabiosis as employed in our laboratory, established a cross-circulation between a Walker carcinosarcoma bearing rat and a normal one (carotid artery to jugular vein, providing for a blood exchange of about 5 ml./min). These workers noted that resulting from cross-circulation for a few hours, the histidine decarboxylase activity of the liver of the normal rat was elevated, although not to the high level existing in the tumour-bearing rat liver, a ratio of elevation similar to that reported by Johnston. The Japanese workers employed a not yet validated modification of the isotopic method described by Kahlson, Rosengren & Thunberg (1963).

In rats bearing this type of tumour morphological changes were not apparent in tissues other than the liver. Nevertheless, the HFC of several organs was found to be elevated in both partners as seen in Table 78. In the skin the HFC was elevated in the tumour-bearing partner only. The absence of a corresponding change in the skin of the non-tumour-bearing counterpart is perplexing considering the readiness with which the skin otherwise generates high HFC. As to the gastric mucosa the figures are inconsistent and not significant in this connexion. It will be recalled that in moribund ascites-tumour bearing mice the HFC was substantially elevated in various tissues even in the absence of apparent morphological changes. Students of the biochemistry of cancer maintain that 'whatever effect the distant tumour produces upon a tissue is largely metabolic, and bears little or no relation to the morphology of the tissues' (Greenstein, 1954).

In this connexion a report by Moolten and Bucher (1967) should be mentioned. They established carotid-to-jugular cross-circulation between partially hepatectomized and normal rats and

Table 78. HFC of tissues of parabiotic rats (ng histamine formed/g tissue)

	Pair 1		Pair 2		Pair 3	
	− Tumour	− Tumour	− Tumour	+ Tumour	− Tumour	+ Tumour
Abdominal skin	61	43	49	989	58	464
Liver	58	44	95	305	278	523
Lung	963	838	1932	3374	3337	7460
Spleen	935	678	1844	2684	4337	4510
Gastric mucosa	15,050	18,500	7139	2149	16,140	21,660
Small intestine	—	—	—	—	—	4
Kidney	6	1	18	35	51	80
Tumour				398		412

− = not detectable.

(Johnston, 1967, *Experientia*, **23**, 152.)

recorded an increased incorporation of ^{14}C-thymidine into hepatic DNA in the normal partner when blood flow was maintained for as long as 19 hours, but found no effect with cross-circulation for 7 hours. From this they suggest that control of liver regeneration resides in a humoral mechanism (blood-borne). They emphasize the well known preferential localization of DNA synthesis and mitosis in regenerating liver to the parts of the lobule that are first exposed to inflowing blood.

Amine formation in the regenerating rat liver

At the start of our work on the HFC of rapidly growing tissues, about ten years ago, high HFC had been discovered in foetal tissues but the studies of wound healing were in an early stage. The growth of the regenerating liver after partial hepatectomy appeared relevant to our study, because, like foetal growth, its rate is very rapid. The remnant grows and reaches the size of the normal liver within 20 days, and the rate of growth is known to be particularly rapid during the first few days after operation. We determined the excretion of histamine and ^{14}C-histamine after ^{14}C-histidine and found increased rates during liver regeneration. The increased excretion subsided when the regenerative growth was completed. In a lecture at the University of Oxford in 1959 on the above, published in 1960, Kahlson hinted that the increased histamine formation may have taken place due to the rapidly growing liver. This paragraph was subordinate to the message of the lecture which concluded with the remarks: 'Through the veil two facts emerge with considerable certainty: 1. Histamine, as linked with the physiological events at present being studied, is not formed from mast cells. 2. As far as physiology is concerned, studies of the rate of histamine formation are more promising than the measurement of histamine content.'

After the high HFC of the tissues of healing skin wounds had been fully established it became apparent that the increased histamine excretion following the operation of partial hepatectomy may in part result from histamine formed in the wounds thereby inflicted. We turned to examining histamine formation in the regenerating liver tissue *in vitro*, employing the sensitive isotopic method. Much to our surprise, the HFC of the regenerating tissue was not significantly higher than in the intact normal liver. In the regenerating liver a higher over-all activity of the amino acid activating enzymes had been reported by Decken and Hultin (1958). We suspected that our failure to find what we were searching for might be due to our ignorance of the optimum conditions for histidine decarboxylase activity in this particular tissue. In the meantime it was discovered that the regenerating rat liver produces the amines spermidine and putrescine at increased rates (Raina, Jänne & Siimes, 1965, 1966; Jänne, 1967). Similar results were obtained by Russell and Snyder (1968) who in the regenerating rat liver found ornithine decarboxylase activity, which produces putrescine, about nine times greater than in sham-operated animals. The problem of these latter amines as related to rapid tissue growth will be discussed later.

Claims for non-validity of our hypothesis regarding HFC and growth

Following G. B. West and his co-workers, Shephard and his colleagues have taken a similar interest in attempts to confirm our hypothesis concerning high HFC in certain rapidly growing tissues. Their endeavour has led them to refute a claim which has never been made, namely that a high HFC exists in all rapidly growing tissues. We have suggested, and still maintain, that in certain rapidly growing tissues there is an association between rapid growth and high HFC. We have further suggested, mainly from observations on healing skin wounds, that this association may have physiological significance, and that this coincidence may constitute an essential circumstance in the process of rapid growth in some tissues. Ever since 1958, we have reported instances of relatively low HFC in some rapidly growing tissues, e.g. muscle and skin in the rat foetus (Kahlson, Rosengren, Westling & White, 1958) and the hamster foetus, in which the HFC is only a small fraction of that in the placenta (Rosengren, 1965). At an early stage of this particular work we suggested that a connexion may exist between high HFC and the anabolic processes of development, growth, and repair (Kahlson, 1960). A connexion of this kind has explicitly been emphasized by Rosengren (1963) with the remark, 'Further exploration is likely to trace connexions with metabolic changes.' The validity of this prediction will be set out in Chapter 22.

Results opposing those obtained in our laboratory have been recorded with methods which are now recognized to be not particularly suitable for the purpose. It may be assumed therefore that some of the conclusions drawn by opponents appear debatable.

Under the title *Histidine Decarboxylase and Growth* Mackay, Reid and Shephard (1961) investigated this problem applying Waton's non-isotopic method. With this method the rate of histamine formation was recorded as being low in liver and kidney of foetal, young and adult guinea-pigs, in extracts of regenerating rays of starfish (*Asterias rubens*), and in chick embryos. It is now known that the method and the pH as employed in this study determines mainly DOPA-decarboxylase, and not histidine decarboxylase. Kameswaran and West (1962), applying the same method as Mackay *et al.* (1961), recorded that the guinea-pig foetus is capable of producing large amounts of histamine (presumably by the agency of DOPA-decarboxylase, since benzene was required in their determinations). In our laboratory the guinea-pig foetus, at a certain stage of development, displayed a HFC of considerable magnitude, engendered by histidine decarboxylase (unpublished).

Shephard and Woodcock (1968), employing Schayer's isotopic method as adapted for use in our laboratory, investigated the histidine decarboxylase activity *in vitro* of a number of hepatomas transplanted into Wistar rats. The hepatomas examined had approximately the same rate of growth. The values obtained are seen in Table 79. With regard to the hepatomas examined, the authors are not impressed by the occurrence of exceedingly high HFC values in a considerable number of the hepatomas, but by the lack of definite correlation between the rate of growth of the tumours and their HFC. The authors conclude that their results

Table 79. Histidine decarboxylase (H-D) activities of normal, foetal, regenerating and malignant liver tissue of the rat

Tissue	Tumour inducer	Source	Rat strain	H-D activity* pH 8.5 with benzene	H-D activity* pH 8.5 without benzene	H-D activity* pH 6.5 without benzene	H-D activity† pH 6.5 with benzene	H-D activity† pH 7.4 without benzene	Histamine in blank incubation (μg/g)
Foetal liver (19 ±1 day gestn.)	—	—	Wistar	31·0	29·5	176·4	274,500	1,092,000	24·3
F-Hep	FDAB	Chester-Beatty	August	28·7	35·5	170·5	183,700	822,000	24·5
Primary liver tumour	DENA	This laboratory	Wistar	10·9	11·7	40·9	16,400	111,200	7·3
D/31 hep.	DAB	University of Nottingham	S–Dawley	1·4	2·0	5·0	3300	9120	6·9
D/30 hep.	DAB	University of Nottingham	S–Dawley	nil	nil	3·5	4990	22,380	3·5
Novikoff hep.	DAB	Inst. du Cancer de Montreal	S–Dawley	0·2	1·0	2·3	580	7190	1·4
P/26 hep.	DAB	Chester-Beatty	Chester-Beatty	1·0	nil	nil	740	7080	nil
U/B/P17	Ethionine	Chester–Beatty	Chester-Beatty	4·2	nil	nil	236	182	nil
Regenerating liver (29 hr)	—	—	Wistar	26·4	2·0	2·0	2016	380	2·0
Normal liver	—	—	Wistar	22·1	1·6	0·4	2080	480	0·7

DAB = 4-Dimethylaminoazobenzene; FDAB = 4′-Fluoro-4-dimethylaminoazobenzene; DENA = Diethylnitrosamine.
* Histamine formed μg/g wet weight.3 hr.   † ^{14}C-Histamine formed dpm/g wet weight.3 hr.
Results are the mean of duplicate determinations on pooled tissues from 3 animals except for foetal liver (pooled livers from a litter of nine) and the primary tumour (one case). (Shephard & Woodcock, 1968, Biochem. Pharmacol. 17, 23.)

do not support the view that histidine decarboxylase has an essential role in regard to rapid tissue growth. The figures of Table 79 rather show that the growth of the hepatomas examined, except for one specimen, is associated with high HFC. The hepatomas investigated by Shephard and Woodcock are cytologically and presumably biochemically different in nature. Notwithstanding, these hepatomas, except for one, have a common feature, the high HFC. In attempts to characterize the enzyme(s) involved, these authors resorted to a peculiar aid, benzene and changing the pH, instead of the differential inhibitors α-methylhistidine and α-methyl-DOPA, described in Chapters 6 and 8, which to an increasing extent are made use of by students of the pertinent enzymes.

Håkanson (1966) has repeated various of our experiments regarding histamine formation. With the method employed, he could confirm the existence of histamine formation in the gastric mucosa, but on three homogenized experimental tumours investigated by him, the Rous rat sarcoma, the Walker rat mammary carcinosarcoma 256, and a hamster melanoma, the two former, but not the melanoma, displayed histamine formation. These results indicated to Håkanson 'that a high histamine-forming capacity is not a necessary feature of tumour growth'. This pronouncement, based on results with three tissues, with a method different from ours, as well as similar statements, appear superfluous in the face of our early conclusion: 'It would be rash to refer to nascent histamine as a factor essential to tissue growth in general' (Kahlson, 1962).

Shephard and Woodcock (1968), besides the hepatomas, investigated the HFC of rat liver regenerating after partial hepatectomy. They found the histidine decarboxylase activity of the regenerating liver significantly lower than that of the corresponding control during the first 48 hr following the operation. This finding was taken as additional indication that a high HFC is not essential for the growth of various types of liver tissue. As reported earlier, it appears to be well established now that the regenerating liver forms amines at a high rate, but they are not histamine, and to these other amines we now turn.

Amines other than histamine in rapid tissue growth

Reference has already been made to the work of the Finnish

group who discovered that in the regenerating rat liver after partial hepatectomy, the synthesis of putrescine from ornithine and arginine was greatly increased immediately after the operation, reaching a maximum at 8 hr (Jänne, 1967). They also disclosed an increased synthesis of spermidine in the regenerating liver (Raina, Jänne & Siimes, 1966; Jänne & Raina, 1968). At about the same time it was reported that the assay for uridine monophosphate by combined ornithine 5′-monophosphate pyrophosphorylase and decarboxylase was enhanced as early as 2 hours after partial hepatectomy, and that this increase was prevented by the administration of actinomycin D (Bresnick, 1965).

Investigations of amine synthesis in rapidly growing tissues were resumed by Russell and Snyder (1968). These examined the activities of ornithine decarboxylase, histidine decarboxylase, and other amino acid decarboxylases in regenerating rat liver, chick embryo, and several rat tumours. In the liver remnant after hepatectomy the ornithine decarboxylase activity was found to be ten times the normal value at 4 hr, and 17 times greater at 16 hr. Liver remnants were assayed for ornithine decarboxylase activity by measuring both $^{14}CO_2$ production and ^{14}C-putrescine formation, and the activity was found to be the same with either method. Fig. 82 from the report of these authors is included because of the strikingly high levels of phenylalanine decarboxylation which they ascribe as presumably produced by the enzyme named 'nonspecific aromatic L-amino acid decarboxylase' by Lovenberg, Weissbach and Udenfriend (1962). Although histidine is not a principal substrate for this latter enzyme, it is capable of decarboxylating histidine. These facts provide little guidance in resolving the problem why the histamine formation is unaltered in the liver remnant (Fig. 82). Shephard and Woodcock (1968), by contrast, reported that in the liver remnant the histidine decarboxylase activity was significantly lowered.

The synthesis of amines and the pattern of histidine and ornithine decarboxylases in the developing chick embryo have been investigated by Dzodzoe and Rosengren (1971). They found histidine decarboxylase activity present mainly in the supernatant fraction of the tissue homogenate, and the activity was relatively low. By contrast, the ornithine decarboxylase activity was high, particularly at the early stage of development; as a noteworthy feature they found that the topographical location of this latter

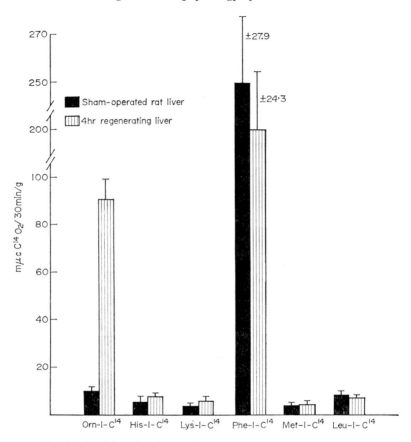

FIG. 82. Activity of amino acid decarboxylases in regenerating rat liver. Vertical bars indicate the SEM for groups of five to ten rats (Russell & Snyder, 1968, *Proc. natn. Acad. Sci. U.S.A.* **60**, 1420).

enzyme changed with the age of the embryo. The patterns of enzyme activity are shown in Fig. 83.

Enzymic changes occurring in regenerating rat liver, entirely different in kind, should be mentioned in passing. Bengmark, Engevik and Olsson (1968) noted after partial hepatectomy an approximately 30-fold increase in ornithine carbamoyl transferase activity in blood serum, indicative of a release of this enzyme from the liver.

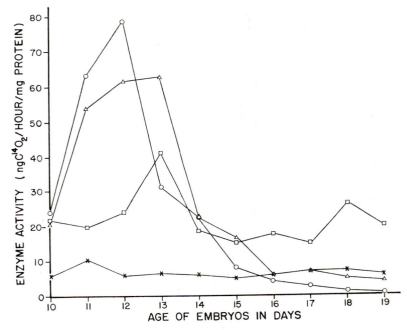

FIG. 83. Patterns of decarboxylase activity in the developing chick embryo: Ornithine decarboxylase: homogenate △—△, supernatant ◯—◯, residue ☐—☐. Histidine decarboxylase (supernatant only): x—x (Dzodzoe & Rosengren, 1971, *Br. J. Pharmac.*, **41**, 294).

Russell and Snyder (1968) investigated the following rat tumours for their content of decarboxylase activities: Morris hepatoma, sarcomas STAT 1 and 3, NRF-TFF sarcoma. For brevity the results are shown in Fig. 84 of their report. Of four different tumours investigated, three exhibited high histidine decarboxylase activity in varying degree: the hepatoma, STAT-3 sarcoma, and NRF-TFF sarcoma; the ornithine decarboxylase activity of these tumours was low. The STAT-1 sarcoma displayed exceedingly high ornithine decarboxylase, comparable to the activity in regenerating rat liver, whereas the histidine decarboxylase activity was very low in this tumour only. The authors emphasize the inverse relationship between the activity of histidine decarboxylase and that of ornithine decarboxylase.

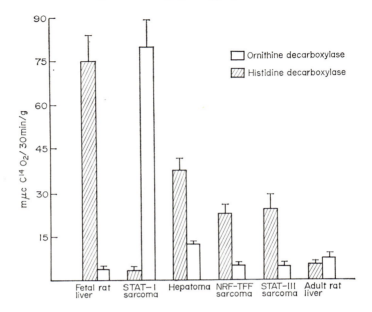

FIG. 84. Ornithine decarboxylase and histidine decarboxylase activity in foetal and adult rat liver and in rat tumours. Tumour columns represent the mean and SEM for two to three animals. The liver data represent the mean and SEM for five to ten animals (Russell & Snyder, 1968, *Proc. natn. Acad. Sci. U.S.A.*, **60**, 1420).

The product of ornithine decarboxylase is putrescine, a diamine, as is histamine. Russell and Snyder assume, on the evidence obtained in the regenerating liver, and in a particular tumour tissue, that in these rapidly growing tissues, putrescine may have a function similar to that of nascent histamine which latter is involved in a great variety of rapidly growing tissues.

On the assumption that in certain rapidly growing tissues putrescine in fact plays a role similar to that assumed for nascent histamine, Kahlson and Rosengren (1970) suggested that the original hypothesis of 1960 regarding HFC and tissue growth should be broadened to state that a high rate of intracellular diamine formation is associated with and presumably essential to certain types of rapid tissue growth.

22

RETARDATION OF PROTEIN SYNTHESIS IN NORMAL AND MALIGNANT TISSUES ON INHIBITING HISTAMINE FORMATION

WHEN we embarked upon experiments to show, as directly as possible, an association between HFC and the rate of protein synthesis, some pertinent information was already available. (1) It had been shown that the formation of collagen, which is a protein, was enhanced during the state of experimentally increased histamine formation. (2) Granulation tissue of a skin wound, which is mainly collagen, generates high histidine decarboxylase activity with a maximum around the 5th day of healing (Fig. 66), the day when the gap of the wound becomes closed. (3) The mammary gland is specifically designed to manufacture protein. Our colleagues Lilja and Svensson (1967) noted that the lactating gland evolves a substantial elevation of HFC and that the increased histamine formation persists in the gland for the whole period of lactation.

If nascent histamine were an essential factor in the process of protein synthesis in certain tissues, then inhibiting histamine formation should retard protein synthesis. This is in fact the case and has been ascertained by the following experiments (Grahn, Hughes, Kahlson & Rosengren, 1969).

The normal tissue of choice is the foetal rat liver which is endowed with histidine decarboxylase activity of about 1000 times that of the young. The rate of protein synthesis was determined by measuring the rate of incorporation of ^{14}C-labelled leucine into liver slices. The incorporation was determined in two fractions of protein referred to respectively as soluble protein fraction and structural protein fraction. The specific radioactivity in labelled proteins was counted in a scintillation spectrometer and expressed as counts/min per milligram protein. The protein mass was also determined. The considerable number of manipulations and processes involved in this investigation are described in the publication by Grahn *et al.* (1969).

Effect of enzyme inhibitors on histamine formation. The results summarized in Table 80 confirm, as have those of other workers, our observation in 1958 of an exceedingly high HFC of the rat foetal liver. Confirmed, also, is the strong inhibition of histamine formation by α-methylhistidine, a compound which has been described in Chapter 8 as a specific inhibitor of histidine decarboxylase. Puromycin, an inhibitor of protein synthesis, was employed with the view to ascertaining whether the radioactivity specified in Table 81 did in fact correspond to the rate of protein synthesis. The experiments detailed in this table provide, in addition, some information on the lifetime of histidine decarboxylase molecules in foetal liver. Puromycin effected complete inhibition of protein synthesis, whereas the histidine decarboxylase activity remained unaffected during the incubation time of 90 min, i.e. the lifetime of this enzyme in rat liver slices is at least 90 min. On injecting rats with inhibitors of protein synthesis and determining histidine decarboxylase activity in homogenized gastric mucosa at varying time intervals after injecting the inhibitor, Snyder and Epps (1968) recorded a rapid fall in gastric histidine decarboxylase activity with a half-life of about 2 hours.

Incorporation of [14]*C-leucine under the influence of* α-*methylhistidine.* It will be recalled that in order to retard growth and healing of wounds the HFC must be strongly inhibited; more than 50% inhibition is required. In these experiments α-methylhistidine was employed in concentrations which inhibited the rate of histamine formation to 15–20% of the control value (Table 80). In experiments, the results of which are summarized in Table 81, the rate of leucine incorporation was reduced to, respectively, 53 and 70% in the soluble and structural protein fractions in foetal liver. By contrast, in the liver of the young, in which the HFC is low, the leucine incorporation into the two types of protein was only insignificantly altered.

Table 80. Formation of ^{14}C-histamine from ^{14}C-histidine in rat liver slices at two age stages during 90 min incubation periods. Figures are mean values expressed as ng histamine per gram tissue $\pm$ S.E. of mean. Figures in parentheses are the number of observations. The levels of significance are referred to control values. P values greater than 0·1 are considered not significant (N.S.)

Tested compounds in final concentrations	19-day rat foetal liver histamine formation		5-day rat young liver histamine formation	
	ng/g tissue	Relative activity (%)	ng/g tissue	Relative activity (%)
Control	$46{,}800 \pm 2300$ (8)	100	$50{\cdot}8 \pm 4{\cdot}1$ (8)	100
DL-α-methyl-histidine 2·5 mM	6900 ± 800 (4) $t = 12{\cdot}14$ P < 0·001	15	$10{\cdot}4 \pm 2{\cdot}8$ (8) $t = 8{\cdot}178$ P < 0·001	20
Puromycin dihydrochloride, 200 μg/ml	$49{,}200 \pm 5200$ (5) $t = 0{\cdot}498$ N.S.	105	$52{\cdot}4 \pm 3{\cdot}7$ (5) $t = 0{\cdot}276$ N.S.	103

(Grahn *et al.* 1969, *J. Physiol.* **200**, 677.)

Extracellular histamine devoid of the action of nascent histamine. In these experiments histamine was added to the incubation mixture in order to see whether extracellular histamine in concentrations considered reasonably physiological could counteract the inhibitory influence on protein synthesis which occurred on the inhibition of histamine formation. Aminoguanidine (AMG) was employed to prevent the oxidative deamination of the histamine added. Foetal liver only was investigated. On adding histamine the inhibitory influence of α-methylhistidine on protein synthesis was slightly less, but the difference appears uncertain (Table 81).

Influence of added histidine. Adding L-histidine to incubated slices of foetal or young liver did not significantly affect the leucine incorporation into protein (Table 81). This may be taken to indicate that the intracellular concentration of the substrate for histidine decarboxylase was optimal under the experimental conditions.

Retardation of protein synthesis in tumours on inhibiting histamine formation. Two rat tumours, the Walker 256 mammary carcino-

Table 81. Incorporation of ^{14}C-L-leucine into rat liver slices at two age stages during 90 min incubation period. Figures are mean values ±S.E. of mean. Figures in parentheses are the number of observations. The levels of significance are referred to control values. P values greater than 0·1 are considered not significant (N.S.)

Tested compounds in final concentrations	19-day rat foetal liver				5-day rat young liver	
	Soluble protein fraction		Structural protein fraction		Soluble protein fraction	Structural protein fraction
	Counts/min ×10³ per mg protein	Relative activity (%)	Counts/min ×10³ per mg protein	Relative activity (%)	Counts/min ×10³ per mg protein	Counts/min ×10³ per mg protein
Control	30·04 ±1·46 (10)	100	19·96 ±1·15 (10)	100	10·21 ±1·34 (8)	6·76 ±1·02 (8)
DL-α-methylhistidine 2·5 mM	15·91 ±0·47 (9) $t=8·778$ $P<0·001$	53	14·16 ±0·55 (9) $t=4·387$ $P<0·001$	71	8·76 ±1·09 (8) $t=0·839$ N.S.	6·18 ±0·95 (8) $t=0·417$ N.S.
Histamine 7·5 µg base/ml, AMG 10^{-4} M, DL-α-methylhistidine 2·5 mM	18·83 ±1·04 (7) $t=5·707$ $P<0·001$	63	16·81 ±1·03 (7) $t=1·934$ $0·05<P<0·1$	84	—	—
L-histidine 3·0 mM	29·68 ±1·43 (5) $t=0·155$ N.S.	99	20·32 ±1·51 (5) $t=0·185$ N.S.	102	7·65 ±1·55 (8) $t=1·247$ N.S.	5·15 ±1·02 (8) $t=1·121$ N.S.
Puromycin dihydrochloride 200 µg/ml.	0·41 ±0·03 (4)	1·4	0·45 ±0·02 (4)	2·3	0·07 ±0·01 (4)	0·11 ±0·02 (4)

(Grahn et al, 1969, J. Physiol. **200**, 677.)

sarcoma and the Rous virus sarcoma, were investigated by Grahn and Rosengren (1970) in a similar way to that described for the foetal rat liver. Both tumours were transplanted subcutaneously to female Sprague–Dawley rats. The Walker 256 mammary tumour was of a local strain, not of Chester Beatty Institute origin, and the Rous virus sarcoma was the same as the one studied earlier. In this particular Walker tumour the HFC was lower than in the Chester Beatty strain which was no longer available for study. It is clear from investigations described previously that in the Walker tumour the HFC is widely different in different strains. The Rous sarcoma displayed a higher HFC (Table 82).

Table 82. Formation of ^{14}C-histamine from 2-ring-^{14}C-L-histidine by rat tumour tissue in 90 min

Walker 256 mammary carcinoma No inhibitor	Inhibitor DL-α-methylhistidine 2·5 mM	
161	55	
54	69	
74	20	
205	210	
134	28	

Rous virus sarcoma No inhibitor	Inhibitor DL-α-methyl-histidine 2·5 mM	NSD-1055 0·5 mM
304	55	
167	47	
154	27	
202	29	
286	33	17
408		12
227	53	20

The figures are ng histamine formed/g wet tissue and are the means of duplicate determinations on samples of minced tissue. Molarities are final concentrations.

(Grahn & Rosengren, 1970, *Experientia*, **26**, 125.)

To inhibit histamine formation, α-methylhistidine was used in both kinds of tumour, and in some experiments on the Rous sarcoma NSD-1055 was also employed. In the Rous sarcoma both α-methylhistidine and NSD-1055 strongly inhibited histamine formation. In the Walker tumour inhibition by α-methylhistidine was inconsistent, and in two cases it had no effect, which indicates absence of histidine decarboxylase in the two cases (Table 82). Some specimens and particular strains of this tumour presumably contain DOPA-decarboxylase, or some other decarboxylase different from histidine decarboxylase, a supposition which has so far not been tested.

The rate of protein synthesis was determined in about 100 mg of minced tumour tissue to which 0·05 μmoles of ^{14}C-leucine (25 mCi per mmole) was added after preincubation for 20 min at 37°C with the inhibitor. Leucine incorporation was about the same in the two protein fractions in both tumours (Table 83). This result differs from that in the rat liver in which the incorporation rate into the insoluble protein fraction was only 66% of that into the soluble fraction. In the two types of tumour, in the absence of inhibitor, the rate of incorporation was nearly the same. In the Walker tumor, leucine incorporation under the influence of α-methylhistidine (2·5mM) was only slightly inhibited, about 20% inhibition in both protein fractions, and in two instances there was no detectable inhibition (Table 83). As indicated above, the occurrence of histidine decarboxylase is inconsistent in this particular tumour, and it would appear that this tumour depends only to a minor extent, or not at all, on nascent histamine for its growth. In the Rous tumour, on the other hand, either of the inhibitors at the concentrations employed depressed leucine incorporation into both protein fractions by about 50%. In this tumour the results are in accord with the observations in rat foetal liver.

Possible mode of action of nascent histamine

Although in some instances so far investigated nascent histamine seems to be essential to protein synthesis, as is pituitary growth hormone, nascent histamine (NHi), obviously, is not a growth hormone. Functionally potent NHi can neither be extracted nor injected. NHi rather constitutes a particular state of high rate of histamine formation. Its action cannot be reproduced by increasing the concentration of extracellular histamine. Its importance becomes evident on experimentally inhibiting or elevating histidine decarboxylase activity. Metaphorically, NHi represents an occurrence, an event.

Table 83. Incorporation of 1-^{14}C-L-leucine into rat tumour tissue in 90 min

Walker 256 mammary carcinoma

Soluble fraction		Insoluble fraction	
No inhibitor	Inhibitor DL-a-Methyl-histidine 2·5 mM	No inhibitor	Inhibitor DL-a-methyl-histidine 2·5 mM
7·2 (3)	5·6 (2)	8·1 (3)	6·3 (2)
7·1 (2)	5·5 (3)	9·1 (2)	9·6 (3)
16·1 (5)	15·1 (3)	28·5 (5)	22·1 (4)
8·2 (3)	5·6 (3)	12·1 (3)	7·1 (3)
10·4 (3)	9·0 (3)	14·6 (3)	11·9 (3)
19·3 (4)	16·1 (4)	24·7 (4)	20·4 (4)

Rous virus sarcoma

Soluble fraction			Insoluble fraction		
No inhibitor	Inhibitor DL-a-methyl-histidine 2·5 mM	NSD-1055 0·5 mM	No inhibitor	Inhibitor DL-a-methyl-histidine 2·5 mM	NSD-1055 0·5 mM
10·7 (3)	4·7 (3)		7·0 (3)	3·3 (3)	
13·8 (4)	8·1 (4)		11·9 (4)	7·9 (4)	
10·1 (5)	4·4 (5)		14·0 (5)	6·3 (5)	
8·7 (2)	4·4 (2)	4·8 (2)	10·0 (2)	5·7 (2)	6·3 (2)
9·6 (2)		5·6 (4)	11·4 (2)		6·7 (4)
7·5 (2)	4·1 (2)	4·3 (4)	9·7 (2)	5·7 (2)	5·1 (4)

The figures are mean counts/min × 10³ per mg protein in samples of minced tissue. The number of determinations in each experiment is given in parenthesis. Molarities are final concentrations.

(Grahn & Rosengren, 1970, *Experientia*, **26**, 125.)

Further work on the mode of action of NHi is likely to investigate more closely its role in protein synthesis. This will involve the search for relationships with nucleic acid, especially RNA. It has been shown by Jänne and Raina (1968) that polyamine synthesis is accelerated under the conditions where RNA is accumulated, e.g. in the regenerating rat liver (see also Raina & Jänne, 1970). The biosynthetic pathway of polyamines generally has been reviewed by Tabor and Tabor (1964), and in the rat liver by Jänne (1967). The problem of the mode of action of NHi is at its very opening, progress in the field is likely to be slow, and prospective investigators in this field will be aware that the mode of action of the pituitary growth hormone has been studied ardently for half a century.

23

THE MAST CELL:
AN UNCONQUERED CHALLENGE

THE discovery of the presence of large amounts of histamine in tissue mast cells, the high mast-cell content of many tissues, and the finding that virtually all the blood histamine is present within the basophils, the 'blood mast cells', led to the belief that the major part of the histamine of the body resides in these cells. The pharmacological actions of injected histamine, especially on the smooth muscle of the arterioli, made it appear likely that the tissue mast cells are involved in physiological processes by the release of histamine and other agents whereby, amongst others, the terminal blood vessels would be adjusted to the needs of the tissues, chiefly during trauma or stress.

Reasonable as these assumptions appeared at the time when they were favoured, they have not been substantiated by experimental evidence. Nor has the proportion of the total body histamine contained in mast cells been established. This amount cannot at the present be determined because no means are known whereby the body *in vivo* can be completely depleted of the histamine contained in mast cells. Even knowing this amount would not add to the understanding of the physiological significance of the mast-cell histamine.

No evidence has been brought forth to show that histamine is released from mast cells under physiological conditions. Indeed, it is difficult to envisage any useful purpose of histamine release from mast cells. Rather, such release would constitute a nuisance, unless catabolism and excretion keep pace with the release, which is not the case in the outburst of histamine release in the state of anaphylaxis. There is no evidence on record that the importance of the mast cells in physiological events is recognizably connected with their content of histamine. The possibility of a role for histamine and heparin in the structural set-up within the mammalian mast cell appears worthy of more vivid exploration. The mast cells of lower vertebrates are reported to be structurally different in that they are devoid of histamine (Reite, 1965). The amount of histamine in mast cells and basophiles gradually increases in reptiles, birds and mammals (for references see Takaya, 1969).

Mast cells in mammals, like other tissue components, form histamine. Schayer (1956a) made the original discovery that mast cells collected from the rat peritoneal cavity formed histamine, i.e. converted ^{14}C-histidine to ^{14}C-histamine. He inferred that the histamine formed by rat mast cells was 'bound in stable condition', that exogenous histamine was not bound, and that the soluble histidine decarboxylase activity gave no sharp pH maximum but was close to its optimum at a range between pH 6·5–7·5. Although the mast cells contain much histamine, presumably owing to the long lifetime of the amine formed in them, their HFC is not particularly high compared with some other tissues. In the skin, which is rich in mast cells, the HFC is low (Table 7). In fact, the HFC is so low that on depleting the mast cells of the skin of part of their histamine content, it takes 2–3 weeks to restore the original content.

The mast cells are not unique in retaining their histamine for a long time. It has been disclosed that non-mast cells exist which resemble mast cells in that the histamine formed therein has a very long lifetime. This was first recorded in experiments on the rat lung and intestines (Kahlson, Rosengren & Thunberg, 1963). These non-mast cells differ from the mast cells proper in that the histamine of the rat lung and intestines is resistant to mobilization by 48/80 and that the histamine of the intestine is not released in anaphylaxis (for histamine release, see Mongar & Schild, 1962).

The study of the mast cell, begun in 1877 by Paul Ehrlich, has proceeded with varied objects, expounded in a considerable number of monographs. The longest and, as it seemed, the most persistent wave of reports, those on histamine release, appears now on the decline, although a computerized analysis has predicted that mast-cell research will reach a peak in 1978 (Goffman, 1966).

Investigators at the State University of New York, Buffalo, have approached the mast cell in a new way, by studying the activity and properties of a chymotrypsin-like enzyme in rat mast cells. The American workers blocked the active centre of chymo-trypsin by di-isopropyl fluorophosphate (DFP) labelled with a radioactive isotope and traced the labelled inhibitor to its site of action, i.e. the active centre of the enzyme within the cell. Among others, they determined the number of the organo-phosphate-reactive protease or esterase molecules in individual mast cells and estimated that they would constitute at least 10% of the total dry mass of the rat mast cell, which already includes much hep-arin, about 30% by weight of the granules. This enormous enzyme concentration suggested to the authors a storage for a degradative role in mast-cell function. Serotonin, but not histamine, had a high affinity for the active centre of the enzyme, suggesting a role for serotonin in inhibiting protease activity in the intact mast cells. The authors emphasize that the reaction with isotopic DFP can be used to label mast cells for *in vivo* studies, to follow the cells and their granules and the fate of the enzyme in functional situa-tions (Darzynkiewicz & Barnard, 1967).

Physiologically intact mast cells from the peritoneal fluid of rats can now be isolated in Ficoll (a synthetic polysaccharide) or in bovine serum albumin (Johnson & Moran, 1967). The mammalian mast cell is likely to stay at the fore for a long time, providing enzymologists with a structurally well defined model of which, perhaps, the most parts remain to be unravelled. Meanwhile, the search for the place of histamine in normal physiology is likely to proceed along the non-mast-cell course.

24

CONCLUDING REMARKS

THE picture unfolding from the broad survey of work on the physiology of histamine during the last decade shows a determined and lately nearly universal change from a static to a more dynamic approach. Spectacular changes in the rate of histamine formation, in histidine decarboxylase activity, occur in physiological circumstances, as first revealed in the rat embryo. This discovery stimulated a continuous wave of reports on histidine decarboxylase and the rate of histamine formation in tissues. The discrepancy in results obtained with different methods must appear bewildering to the general reader. Two methods are increasingly recognized as useful and giving consistently reproducible results: the whole-body histamine formation as reflected in the urinary excretion of non-isotopic histamine and its metabolites, and Schayer's isotopic method in its various adaptations. Progress in this field would be furthered and confusion lessened if adequate techniques were generally employed.

In the field which Henry Dale designated as 'autopharmacology', the study of effects exerted by body constituents designed to control specific functions, such as acetylcholine and catecholamines, a lead in exploring their physiological significance was sought by recording the effects produced by the agents when injected. This approach has been of little or no avail in finding the place of histamine in the machinery of the body. The actions of endogenously formed histamine cannot be reproduced by injected, extracellular histamine. This pertains particularly to 'nascent' histamine, the metabolic action of which appears to be linked to the very process of its formation. In this connexion it should be re-emphasized, in passing, that we have made observations indicating that endogenously formed histamine is catabolized in a different way from that injected. This circumstance should urge the further development of non-isotopic methods of determining the principal histamine catabolites.

Students of the physiology of histamine are denied the guidance that would be provided if specific antagonistic compounds were available. The principal actions of this amine, in the ranking as we see them at present, namely the part played in gastric secretion and in metabolic processes of tissue growth and protein synthesis, are not obviated by antihistaminics. This is presumably because in these functions histamine is not engaging receptors of the conventional sort. Some help, although limited in scope, has been found in newly discovered means to inhibit histamine formation *in vivo*. The limitation is imposed, in our belief, by the feed-back relation between histidine decarboxylase and the end-product histamine, a relation which by its very nature is likely to exclude the feasibility of inhibiting histamine formation entirely.

Those familiar with the history of histamine studies will recall that Henry Dale, with his esteem for the work and views of Thomas Lewis, did not, in the end, commit himself to any suggestion of release of tissue histamine, except 'for its contribution to local and general reactions, by which the organism as a whole and its separate tissues respond to various chemical, immunological, or physical assaults upon the integrity of their living cells' (Dale, 1953). Endeavours for decades had not succeeded in demonstrating histamine release in any purely physiological event. Now, at last, by sheer chance, a case of physiological histamine release has been shown to occur, namely a lowering of the gastric mucosal histamine content on feeding a rat deprived of food, a phenomenon which we interpret as the first link in a chain of alterations in the mucosal histamine metabolism designed to stimulate and sustain the secretion of hydrochloric acid.

From our account it would perhaps appear that histamine has been found to be definitely and potentially associated with too many functions to be specifically useful in the machinery of the body. A similar situation applies to the catecholamines, and in part to acetylcholine, whose diversity of functions is well established. Even gastrin, until very recently believed to be specific to one target organ, has been found in experiments by Gregory and his colleagues to exert a variety of extra-gastric actions, when injected.

The group of workers in this laboratory which for many years have been persistently searching for a place for histamine in normal physiology has recorded a number of observations, some by chance, some by deliberately designed experiments. This Monograph is in the main an account of observations made in a wide field of experiments. In retrospect, we trust that our unfading line of observations will not be subjected to Lord Moulton's castigation: 'When we are reduced to observation, Science crawls.'

We have, with caution, also attempted somewhat loftier objectives in that we have endeavoured to amalgamate otherwise perhaps trivial observations into conceptions and theories. Our conception of the physiological significance of histamine formation and its dynamic state stands as generally accepted ground in current histamine research. A solid body of observations provides the basis of theories which causally associates gastric mucosal histamine release and formation with excitation of the parietal cells to secrete and, 'nascent' histamine with the anabolic processes of growth and protein synthesis in certain tissues. Furthermore, we are airing views and ideas in connexion with observations which must be carried further, e.g. intravascular histamine formation as a factor in vasodilatation, in order to justify definite statements and conceptions.

In conclusion, regarding theories, we quote a passage of a report by Bertrand Russell (1968), to the Council of Trinity College in 1930, the reading of which gave us great pleasure. 'The theories contained in this new work of Wittgenstein's are novel, very original, and indubitably important. Whether they are true, I do not know. As a logician who likes simplicity, I should wish to think that they are not, but from what I have read of them I am quite sure that he ought to have an opportunity to work them out, since when completed they may easily prove to constitute a whole new philosophy.' As to our own theories we tread on humbler grounds.

Still, our theories have the merit of being applicable to the test of experiment. An increasing number of colleagues are employing their skill and critical minds along the lines of these concepts and theories, refuting, corroborating or sharpening them, so in the end, which appears distant, the issues at stake will be worked out and, as we hope, never completed, for the sheer fascination of the adventures.

L

REFERENCES

Ackermann, D. (1910). Uber den bakteriellen Abbau des Histidins. *Hoppe-Seyler's Z.physiol. chem.* **65**, 504–510.

Adam, H. M. (1950). Excretion of histamine in human urine. *Q. Jl exp. Physiol.* **35**, 281–293.

— (1961). Histamine in the central nervous system and hypophysis of the dog. In *Regional Neurochemistry*, edited by Kety, S. S. & Elkes, J., pp. 293–306. London: Pergamon Press.

Adam, H. M., Card, W. I., Riddell, M. J., Roberts, M. & Strong, J. A. (1954). The effect of intravenous infusions of histamine on the urinary histamine and on gastric secretion in man. *Br. J. Pharmac.* **9**, 62–67.

Adam, H. M., Hardwick, D. C. & Spencer, K. E. V. (1957). A method of estimating histamine in plasma. *Br. J. Pharmac.* **12**, 397–405.

Adam, H. M. & Hye, H. K. A. (1964). Effect of drugs on the concentration of histamine in the hypophysis and proximate parts of the brain. *J. Physiol.* **171**, 37–38P.

— (1966). Concentration of histamine in different parts of brain and hypophysis of cat and its modification by drugs. *Br. J. Pharmac.* **28**, 137–152.

Adam, H. M., Hye, H. K. A. & Waton, N. G. (1964). Studies on uptake and formation of histamine by hypophysis and hypothalamus in the cat. *J. Physiol.* **175**, 70–71P.

Adashek, K. & Grossman, M. I. (1963). Response of rats with gastric fistulas to injection of gastrin. *Proc. Soc. exp. Biol. Med.* **112**, 629–631.

Ahlmark, A. (1944). Studies on the histaminolytic power of plasma with special reference to pregnancy. *Acta physiol. scand.* **9**, suppl. 28, 5–107.

Ahlström, C. G., Johnston, M. & Kahlson, G. (1966). Histamine formation in tumour-bearing rats. *Life Sci.*, Oxford **5**, 1633–1640.

Alivisatos, S. G. A. (1966). Histamine–pyridine coenzyme interactions. In *Handbook of Experimental Pharmacology*, edited by Eichler, O. & Farah, A., vol. 18/1, pp. 726–733. Berlin: Springer.

Amin, A. H., Crawford, T. B. B. & Gaddum, H. J. (1954). The distribution of substance P and 5-hydroxytryptamine in the central nervous system of the dog. *J. Physiol.* **126**, 596–618.

Angervall, L., Enerbäck, L. & Westling, H. (1960). Urinary excretion of 5-hydroxy-indoleacetic acid and histamine in the pregnant rat. *Experientia* **16**, 209–210.

Annau, E., Manginelli, A. & Roth, A. (1951). Increased weight and mitotic activity in the liver of tumour-bearing rats and mice. *Cancer Res.* **11**, 304–307.

Anrep, G. V., Ayadi, M. S., Barsoum, G. S., Smith, J. R & Talaat, M. M. (1944). The excretion of histamine in urine. *J. Physiol.* **103**, 155–174.

Anrep, G. V., Barsoum, G. S. & Ibrahim, A. (1947). The histaminolytic action of blood during pregnancy. *J. Physiol.* **106**, 379–393.

Ardill, B. L., Cumming, J. D., Fentem, P. H. & Holton, P. (1969). The absence of effect of gastrin II on skin and voluntary muscle blood flow in a human subject. *J. Physiol.* **203**, 57–59P.

Arnoldsson, H., Helander, C. G., Helander, E., Lindell, S. E., Lindholm, B., Olsson, O., Roos, B. E., Svanborg, A., Soderholm, B. & Westling, H. (1962). Elimination of C^{14}-histamine from the blood in man. *Scand. J. clin. Lab. Invest.* **14**, 241–246.

Arunlakshana, O., Mongar, J. L. & Schild, H. O. (1954). Potentiation of pharmacological effects of histamine by histaminase inhibitors. *J. Physiol.* **123**, 32–54.

Ash, A. S. F. & Schild, H. O. (1966). Receptors mediating some actions of histamine. *Br. J. Pharmac.* **27**, 427–439.

Aures, D. & Clark, W. G. (1964). A rotating diffusion chamber for $C^{14}O_2$ determination as applied to inhibitor studies on mouse mast cell tumour histidine decarboxylase. *Analyt. Biochem.* **9**, 35–47.

Aures, D., Winqvist, G. & Hansson, E. (1965). Histamine formation in the blood and bone marrow of the guinea-pig. *Am. J. Physiol.* **208**, 186–189.

Babkin, B. P. (1938). The abnormal functioning of the gastric secretory mechanism as a possible factor in the pathogenesis of peptic ulcer. *Can. med. Ass. J.* **38**, 421–429.

— (1950). *Secretory Mechanism of the Digestive Glands* (2nd ed.). New York: Hoeber.

Banik, U. K., Kobayashi, Y. & Ketchel, M. M. (1963). Effect of aminoguanidine, histamine, histidine and compound 48/80 on pregnancy in rats and mice. *J. Reprod. Fertil.* **6**, 179–182.

Barcroft, H. (1963). Circulation in skeletal muscle. In *Handbook of Physiology*, Section 2, Circulation, vol. 2, pp. 1353–1385.

Barger, G. & Dale, H. H. (1910). Chemical structure and sympathomimetic action of amines. *J. Physiol.* **41**, 19–59.

Bartosch, R., Feldberg, W. & Nagel, E. (1932). Das Freiwerden eines histaminähnlichen Stoffes bei der Anaphylaxie des Meerschweinchens. *Pflügers Arch. ges. Physiol.* **230**, 129–153.

Beaver, M. H. & Wostmann, B. S. (1962). Histamine and 5-hydroxytryptamine in the intestinal tract of germ-free animals, animals harbouring one microbial species and conventional animals. *Br. J. Pharmac.* **19**, 385–393.

Beilenson, S., Schachter, M. & Smaje, L. H. (1968). Secretion of kallikrein and its role in vasodilatation in the submaxillary gland. *J. Physiol.* **199**, 303–317.

Bengmark, S., Engevik, L. & Olsson, R. (1968). Changes in ornithine carbamoyl transferase activity in serum and in liver after partial hepatectomy in rats. *Scand. J. Gastroent.* **3**, 264–266.

Bennett, A. (1965). Effect of gastrin on isolated smooth muscle preparations. *Nature, Lond.* **208**, 170–173.

Best, C. H. & McHenry, E. W. (1930). The inactivation of histamine. *J. Physiol.* **70**, 349–372.

Bhattacharya, B. K. & Feldberg, W. (1958). Perfusion of cerebral ventricles: effects of drugs on outflow from the cisterna and the aqueduct. *Br. J. Pharmac.* **13**, 156–162.

Bianchi, R. G. & Cook, D. L. (1968). Pharmacological studies on an antigastrin and antiulcer agent, SC-15396. *Fedn. Proc.* **27**, 1331–1336.

Bjurö, T. (1963). Histamine metabolism in adrenalectomized rats. *Acta physiol. scand.* **60**, suppl. 220, 1–68.

Bjurö, T., Lindberg, S. & Westling, H. (1961). Observations on histamine in pregnancy and the puerperium. *Acta obstet. gynec. scand.* **40**, 152–173.

Bjurö, T., Westling, H. & Wetterqvist, H. (1961). Effect of thyroid hormones on histamine formation in the rat. *Br. J. Pharmac.* **17**, 479–487.

— (1964a). The formation of ^{14}C-histamine *in vivo* in normal rats and in rats treated with liothyronine. *Br. J. Pharmac.* **23**, 433–439.

— (1964b). The effect of cortisone on the metabolism of histamine in hyperthyroid rats. *Int. Archs. Allergy appl. Immun.* **24**, 311–317.

Black, J. W. & Stephenson, J. S. (1962). Pharmacology of a new adrenergic beta-receptor-blocking compound (Nethalide). *Lancet* **ii**, 311–314.

Blair, E. L., Dutt, B., Harper, A. A. & Lake, H. J. (1961). Evidence that gastrin is not a general histamine-releasing agent. *J. Physiol.* **157**, 53–54P.

— (1962). Examination of evidence that histamine is involved in gastrin-stimulated gastric secretion. *Proc. XXII Int. Physiol. Congr.* **2**, Abstract No. 377.

Blaschko, H. & Bonney, R. (1962). Spermine oxidase and benzylamine oxidase. Distribution, development and substrate specificity. *Proc. R. Soc. B.* **156**, 268–279.

Blaschko, H. & Duthie, R. (1944). Diamidines as inhibitors of enzyme action. *Biochem. J.* **38**, XXV.

Blaschko, H., Fastier, F. N. & Wajda, I. (1951). The inhibition of histaminase by amidines. *Biochem. J.* **49**, 250–253.

Born, G. V. R. & Sewing, K. FR. (1967). Histamine in the plasma and gastric juice of cats during infusions of (^{14}C) histamine. *J. Physiol.* **193**, 397–403.

Bresnick, E. (1965). Early changes in pyrimidine biosynthesis after partial hepatectomy. *J. biol. Chem.* **240**, 2550–2556.

Brodie, B. B., Beaven, M. A., Erjavec, F. & Johnson, H. L. (1966). Uptake and release of H³-histamine. In *Mechanisms of Release of Biogenic Amines*, vol. 5, edited by Euler, U. S., Rosell, S. & Uvnäs, B. pp. 401–415. London: Pergamon Press.

Brown, D. D., Tomchick, R. & Axelrod, J. (1959). The distribution and properties of a histamine-methylating enzyme. *J. biol. Chem.* **234**, 2948–2950.

Buffoni, F. (1966). Histaminase and related amine oxidases. *Pharmac. Rev.* **18**, 1163–1199.

Burgus, R., Dunn, T. F., Desiderio, D., Ward, D., Vale, W. & Guillemin, R. (1970). Characterization of ovine hypothalamic hypophysiotropic TSH-releasing factor. *Nature, Lond.* **226**, 321–325.

Burkard, W. P., Gey, K. F. & Pletscher, A. (1963). Diamine oxidase in the brain of vertebrates. *J. Neurochem.* **10**, 183–186.

Burkhalter, A. (1962). The formation of histamine by fetal rat liver. *Biochem. Pharmac.* **11**, 315–322.

Burn, J. H. (1950). A discussion on the action of local hormones. *Proc. Roy. Soc. B.* **137**, 281–320.

Burn, J. H. & Dale, H. H. (1926). The vasodilator action of histamine and its physiological significance. *J. Physiol.* **61**, 185–214.

Caren, J. F., Aures, D. & Johnson, L. R. (1969). Effect of secretin and cholecystokinin on histidine decarboxylase activity in the rat stomach. *Proc. Soc. exp. Biol. Med.* **131**, 1194–1197.

Caridis, D. T., Porter, J. F. & Smith, G. (1968). Detection of histamine in venous blood from the stomach during acid secretion evoked by intravenous pentagastrin. *Lancet* **i**, 1281–1282.

Carlsten, A. (1950a). On the sources of the histaminase present in thoracic duct lymph. *Acta physiol. scand.* **20**, suppl. 70, 5–26.

— (1950b). Effect of adrenalectomy on lymph and plasma histaminase. *Acta physiol. scand.* **20**, suppl. 70, 33–46.

— (1950c). No change in histaminase content of lymph and plasma in cats during pregnancy. *Acta physiol. scand.* **20**, suppl. 70, 27–31.

Carlsten, A., Kahlson, G. & Wicksell, F. (1949a). The strong histaminolytic activity of lymph and its bearing on the distribution of histamine between lymph and plasma in dogs. *Acta physiol. scand.* **17**, 370–383.

Carlsten, A., Kahlson, G. & Wicksell, F. (1949b). Accumulation in the canine lymph during reactive hyperaemia of a substance contracting the guinea-pig's gut which is not histamine. *Acta physiol. scand.* **17**, 384–394.

Carlsten, A. & Wood, D. R. (1951). Increased lymph histaminase in adrenalectomized cats and its restoration by adrenocortical extract but not by adrenaline. *J. Physiol.* **112**, 142–148.

Chain, E. B. (1965). Landmarks and perspectives in biochemical research. *Br. med. J.* **1**, 273–278.

Ciba Symposium on Histamine (1956). Edited by Wolstenholme, G. E. W. & O'Connor, C. M. London: Churchill.

Clark, W. G. (1963). Inhibition of amino acid decarboxylases. In *Metabolic Inhibitors* vol. 1, pp. 315–381, New York: Academic Press.

Clark, C. G., Curnow, V. J., Murray, J. G., Stephens, F. O. & Wyllie, J. H. (1964). Mode of action of histamine in causing gastric secretion in man. *Gut* 5, 537–545.

Clark, D. H. & Tankel, H. I. (1954). Gastric acid and plasma-histaminase during pregnancy. *Lancet* ii, 886–887.

Clark, W. G. & Ungar, G. (1964). Histamine and the nervous system. A symposium. *Fedn Proc.* 23, 1095–1116.

Clementz, B. & Rydberg, C. E. (1949). An ordinate recorder for measuring drop flow. *Acta physiol. scand.* 17, 339–344.

Code, C. F. (1956). Histamine and gastric secretion. In *Histamine* edited by Wolstenholme, G. E. W. & O'Connor, C. M., pp. 189–219. London: Churchill.

— (1964). Histamine in human disease. *Mayo clin. Proc.* 39, 715–737.

Code, C. F., Blackburn, C. M., Livermore, G. R. & Ratke, H. V. (1949). A method for the quantitative determination of gastric secretory inhibition. *Gastroenterology* 13, 573–588.

Code, C. F. & Hallenbeck, G. A. (1961). Effect of pregnancy on the histidine decarboxylase activity of the rat's stomach. *J. Physiol.* 159, 66–67P.

Colfer, H. F., de Groot, J. & Harris, G. W. (1950). Pituitary gland and blood lymphocytes. *J. Physiol.* 111, 328–334.

Colin-Jones, D. G. & Himsworth, R. L. (1970). The location of the chemoreceptor controlling gastric acid secretion during hypoglycaemia. *J. Physiol.* 206, 397–409.

Contractor, S. F. & Jeacock, M. K. (1967). A possible feed-back mechanism controlling the biosynthesis of 5-hydroxytryptamine. *Biochem. Pharmac.* 16, 1981–1987.

Conway, E. J. (1950). *Microdiffusion Analysis and Volumetric Error.* London: Crosby Lockwood.

Cook, D. L. & Bianchi, R. G. (1967). SC-15396: A new antiulcer compound possessing anti-gastrin activity. *Life Sci. Oxford* 6, 1381–1387.

Crean, G. P. (1963). The endocrine system and the stomach. *Vitams Horm.* 21, 215–280.

— (1968). Effect of hypophysectomy on the gastric mucosa of the rat. *Gut* 9, 332–342.

Cumming, J. D., Haigh, A. L., Harries, E. H. L. & Nutt, M. E. (1961). The measurement of total gastric blood flow and its relationship to gastric acid secretion in the anaesthetized dog. *J. Physiol.* 157, 39P.

Cumming, J. D., Harries, E. H. L. & Holton, P. (1965). The effects of gastrin II and histamine on the vascular resistance and oxygen consumption of the stomach in anaesthetized dogs. *J. Physiol.* 181, 33P.

Dahlbäck, O., Hansson, R., Tibbling, G. & Tryding, N. (1968). The effect of heparin on diamine oxidase and lipoprotein lipase in human lymph and blood plasma. *Scand. J. clin. Lab. Invest.* 21, 17–25.

Dale, H. H. (1948). Antihistamine substances. *Brit. J. Med.* II, 281–283.

— (1953). In Introduction chapter to *Adventures in Physiology: with Excursions into Autopharmacology.* London: Pergamon Press.

Dale, H. H. (1962). Introduction remarks at the *Symposium on Histamine Metabolism* at the 22nd congress of physiology in Leiden, September 1962 (Read, but not published).

Dale, H. H. & Feldberg, W. (1934). The chemical transmitter of vagus effects to the stomach. *J. Physiol.* **81**, 320–334.

Danforth, D. N. & Gorham, F. (1937). Placental histaminase. *Am. J. Physiol.* **119**, 294–295.

Darzynkiewicz, Z. & Barnard, E. A. (1967). Specific proteases of the rat mast cell. *Nature, Lond.* **213**, 1198–1202.

Davson, H. (1970). *A Textbook of General Physiology*, London: Churchill.

Decken, v. A. & Hultin, T. (1958). Activity of amino acid activating enzyme systems during liver regeneration. *Acta chem. scand.* **12**, 597.

Demis, J. D. & Brown, D. D. (1961). Histidine metabolism in urticaria pigmentosa. *J. invest. Derm.* **36**, 253–257.

Dempster, W. J. (1957). *An Introduction to Experimental Surgical Studies*. Oxford: Blackwell.

Dixon, J. B. (1959). Histamine, 5-hydroxytryptamine and serum globulins in the foetal and neo-natal rat. *J. Physiol.* **147**, 144–152.

Dixon, M. & Webb, E. C. (1964). *Enzymes*, 2nd ed. London: Longmans.

Dotevall, G., Lindell, S. E. & Westling, H. (1963). Histamine studies in a case of Zollinger–Ellison's syndrom. *Acta med. scand.* **174**, 325–328.

Douglas, W. W., Feldberg, W., Paton, W. D. M. & Schachter, M. (1951). Distribution of histamine and substance P in the wall of the dog's digestive tract. *J. Physiol.* **115**, 163–176.

Dragstedt, C. A. & Gebauer-Fuelnegg, E. (1932). II. The nature of a physiologically active substance appearing during anaphylactic shock. *Am. J. Physiol.* **102**, 520–526.

Duncan, J. G. C. & Waton, N. G. (1968). Absorption of histamine from the gastrointestinal tract of dogs *in vivo*. *J. Physiol.* **198**, 505–515.

— (1969). Further studies on the origin of tissue histamine with special reference to intestinal absorption. *Int. Arch. Allergy appl. Immun.* **36**, 31–40.

Duner, H. & Pernow, B. (1958). Histamine and leukocytes in blood during muscular work in man. *Scand. J. clin. Lab. Invest.* **10**, 394–396.

Duner, H., Pernow, B. & Tribukait, B. (1958). Der Einfluss von Hypoxie auf das Bluthistamin, die Plasma-Protein-Concentration und die Leucocytenzahl des Meerschweinchens. *Acta physiol. scand.* **44**, 365–374.

Dzodzoe, Y. C. G. & Rosengren, E. (1971). Decarboxylases of histidine and ornithine in chick embryo. *Br. J. Pharmac.* **41**, 294–301.

Edery, H. & Lewis, P. (1963). Kinin-forming activity and histamine in lymph after tissue injury. *J. Physiol.* **169**, 568–583.

Ehinger, B. & Thunberg, R. (1967). Induction of fluorescence in histamine containing cells. *Expl. Cell Res.* **47**, 116–122.

Emmelin, N. (1945). Ultrafiltration as a method of preparing blood plasma for quantitative estimation of histamine. *Acta physiol. scand.* **9**, 378–381.

Emmelin, N. (1951). The disappearance of injected histamine from the blood stream. *Acta physiol. scand.* **22,** 379–390.

Emmelin, N. & Feldberg, W. (1947). The mechanism of the sting of the common nettle (*Urticaria urens*). *J. Physiol.* **106,** 440–455.

Emmelin, N. & Henriksson, K. G. (1953). Depressor activity of saliva after section of the chorda tympani. *Acta physiol. scand.* **30,** suppl. 111, 75–82.

Emmelin, N. & Kahlson, G. (1944). Histamine as a physiological excitant of acid gastric secretion. *Acta physiol. scand.* **8,** 289–304.

Emmelin, N., Kahlson, G. & Wicksell, F. (1940). The role of histamine in the cephalic and gastric phases of gastric secretion. Submitted for publication and withdrawn by the authors.

— (1941). Histamine in plasma and methods of its estimation. *Acta physiol. scand.* **2,** 123–142.

Erspamer, V. (1961). Pharmacologically active substances of mammalian origin. *A. Rev. Pharmacol.* **1,** 175–218.

Falck, B., Hillarp. N. Å., Thieme, G. & Torp, A. (1962). Fluorescence of catechol amines and related compounds condensed with formaldehyde. *J. Histochem. Cytochem.* **10,** 348–354.

Fawcett, C. P., Reed, M., Charlton, H. M. & Harris, G. W. (1968). The purification of luteinizing-hormone-releasing factor with some observations on its properties. *Biochem. J.* **106,** 229–236.

Feldberg, W. & Harris, G. W. (1953). Distribution of histamine in the mucosa of the gastro-intestinal tract of the dog. *J. Physiol.* **120,** 352–364.

Feldberg, W. & Schilf, E. (1930). *Histamin.* Berlin: Springer.

Feldberg, W. & Talesnik, J. (1953). Reduction of tissue histamine by compound 48/80. *J. Physiol.* **120,** 550–568.

Ferreira, S. H. Corrado, A. P. & Rocha e Silva, M. (1969). Kinins and vasodilatation of the sub-mandibular salivary gland of dogs. *Abstr. 4th Int. Congr. Pharmac.* pp. 206–207.

Fleisch, A. (1930). Der Pulszeitschreiber; ein Apparat zur Aufzeichnung der zeitlichen Pulsintervalle als Ordinate. *Z. ges. exp. Med.* **72,** 384–400.

Fleisch, A. & Sibul, I. (1933). Uber nutritive Kreislaufregulierung–II. Die Wirkung von pH, intermediären Stoffwechselprodukten und anderen biochemischen Verbindungen. *Pflüger's Arch. ges. Physiol.* **231,** 787–807.

Folkow, B. (1949). Intravascular pressure as a factor regulating the tone of the small vessels. *Acta physiol. scand.* **17,** 289–310.

Folkow, B., Haeger, K. & Kahlson, G. (1948). Observations on reactive hyperaemia as related to histamine, on drugs antagonizing vasodilatation induced by histamine and on vasodilator properties of adenosinetriphosphate. *Acta physiol. scand.* **15,** 264–278.

Forrester, T. & Lind, A. R. (1969). Identification of adenosine triphosphate in human plasma and the concentration in the venous effluent of forearm muscles before, during and after sustained contractions. *J. Physiol.* **204,** 347–364.

Forte, J. G. & Davies, R. E. (1964). Relation between hydrogen ion secretion and oxygen uptake by gastric mucosa. *Am. J. Physiol.* **206,** 218–222.

Fortier, C. (1951). Dual control of adrenocorticotrophin release. *Endocrinology* **49,** 782–788.

Fram, D. H. & Green, J. P. (1965). The presence and measurement of methylhistamine in urine. *J. biol. Chem.* **240,** 2036–2042.

Franco-Browder, S., Masson, G. M. & Corcoran, A. C. (1959). Induction of acute gastric lesions by histamine liberators in rats. *J. Allergy* **30,** 1–10.

Fuche, J. & Kahlson, G. (1957). Histamine as a stimulant to the anterior pituitary gland as judged by the lymphopenic response in normal and hypophysectomized rabbits. *Acta physiol. scand.* **39,** 327–347.

Furano, A. V. & Green, J. P. (1964). The uptake of biogenic amines by mast cells of the rat. *J. Physiol.* **170,** 263–271.

Gaddum, J. H. (1929). An outflow recorder. *J. Physiol.* **67,** 1–2P.

— (1951). The metabolism of histamine. *Br. med. J.* **11,** 987–991.

Ganrot, P. O., Rosengren, A. M. & Rosengren, E. (1961). On the presence of different histidine decarboxylating enzymes in mammalian tissues. *Experientia* **17,** 263–264.

Garden, J. W. (1966). Plasma and sweat histamine concentrations after heat exposure and physical exercise. *J. appl. Physiol.* **21,** 631–635.

Gaskell, W. H. (1880). On the tonicity of the heart and blood vessels. *J. Physiol.* **3,** 48–75.

Gautvik, K. M., Hilton, S. M. & Torres, S. H. (1970). Consumption of kininogen in the submandibular salivary gland when activated by chorda stimulation. *J. Physiol.* **211,** 49–61.

Gavin, G., McHenry, E. W. & Wilson, M. J. (1933). Histamine in canine gastric tissue. *J. Physiol.* **79,** 234–238.

Ghadially, F. N. & Parry, E. W. (1965). Ultrastructure of the liver of the tumour-bearing host. *Cancer* **18,** 485–495.

Goffman, W. (1966). Mathematical approach to the spread of scientific ideas—the history of mast cell research. *Nature, Lond.* **212,** 449–452.

Graham, P., Kahlson, G. & Rosengren, E. (1964). Histamine formation in physical exercise, anoxia and under the influence of adrenaline and related substances. *J. Physiol.* **172,** 174–188.

Graham, P. & Schild, H. O. (1967). Histamine formation in the tuberculin reaction of the rat. *Immunology* **12,** 725–727.

Grahn, B., Hughes, R., Kahlson, G. & Rosengren, E. (1969). Retardation of protein synthesis in rat foetal liver on inhibiting rate of histamine formation. *J. Physiol.* **200.** 677–685.

Grahn, B. & Rosengren, E. (1968). Adrenaline induced acceleration of histamine formation *in vitro*, studied by two isotopic methods. *Br. J. Pharmac.* **33,** 472–479.

— (1970). Retardation of protein synthesis in rat tumours on inhibiting histamine formation. *Experientia* **26,** 125–126.

Grandjean, L. C. (1947). Effect of aneurine on the intestinal contraction produced by histamine. *Acta pharmac. tox.* **3,** 257–260.

Granerus, G. (1968). Urinary excretion of histamine, methylhistamine and methylimidazoleacetic acids in man under standardized dietary conditions. *Scand. J. clin. Lab. Invest.* **22,** suppl. 104, 59–68.

Granerus, G. & Magnusson, R. (1965). A method for semiquantitative determination of 1-methyl-4-imidazoleacetic acid in human urine. *Scand. J. clin. Lab. Invest.* **17,** 483–490.

Granerus, G., Svensson, S. E. & Wetterqvist, H. (1969). Histamine in alcoholic drinks. *Lancet* **i,** 1320.

Granerus, G., Svensson, S. E., Wetterqvist, H. & White, T. (1969). The metabolism of histamine in a case of cold urticaria. *Acta allerg.* **24,** 258–260.

Granerus, G., Wetterqvist, H. & White, T. (1968). Histamine metabolism in healthy subjects before and during treatment with aminoguanidine. *Scand. J. clin. Lab. Invest.* **22,** suppl. 104, 39–48.

Green, H. & Erickson, R. W. (1964). Effect of some drugs upon rat brain histamine content. *Int. J. Neuropharmac.* **3,** 315–320.

Greenstein, J. P. (1954). *Biochemistry of Cancer.* New York: Academic Press.

Gregory, R. A. & Tracy, H. J. (1961). The preparation and properties of gastrin. *J. Physiol.* **156,** 523–543.

— (1963). Constitution and properties of two gastrins extracted from hog antral mucosa. *J. Physiol.* **169,** 18–19P.

— (1964). The constitution and properties of two gastrins extracted from hog antral mucosa. *Gut* **5,** 103–117.

Groot, J. de & Harris, G. W. (1950). Hypothalamic control of the anterior pituitary gland and blood lymphocytes. *J. Physiol.* **111,** 335–346.

Grossman, M. I. (1961). Cholinergic potentiation of the response to gastrin. *J. Physiol.* **157,** 14–15P.

Guggenheim, M. (1912). Zur Kenntnis der Wirkung des *p*-Oxyphenyläthylamins. *Therap. Monasch.* **26,** 795–798.

Guirard, B. M. & Snell, E. E. (1954). Pyridoxal phosphate and metal ions as cofactors for histidine decarboxylase. *J. Am. chem. Soc.* **76,** 4745–4746.

Gustafsson, B., Kahlson, G. & Rosengren, E. (1957). Biogenesis of histamine studied by its distribution and urinary excretion in germ-free reared and not germ-free rats fed a histamine-free diet. *Acta physiol. scand.* **41,** 217–228.

Haartmann, U., Kahlson, G. & Steinhardt, C. (1966). Histamine formation in germinating seeds. *Life Sci. Oxford* **5,** 1–9.

Haddy, F. J. & Scott, J. B. (1968). Metabolically linked vasoactive chemicals in local regulation of blood flow. *Physiol. Rev.* **48,** 688–707.

Haeger, K., Jacobsohn, D. & Kahlson, G. (1952). The levels of histaminase and histamine in the gastro-intestinal mucosa and kidney of the cat deprived of the hypophysis or the adrenal glands. *Acta physiol. scand.* **25,** 243–254.

— (1953*a*). The level of histaminase in thoracic duct lymph as an indicator of adrenocortical activity. *Acta physiol. scand.* **30**, suppl. 111, 170–176.

Haeger, K., Jacobsohn, D. & Kahlson, G. (1953*b*). Atrophy of the gastrointestinal mucosa following hypophysectomy or adrenalectomy. *Acta physiol. scand.* **30**, suppl. 111, 161–169.

Haeger, K. & Kahlson, G. (1952*a*). Distribution of histamine and histaminase in the gastro-intestinal mucosa of fed and starved cats. *Acta physiol. scand.* **25**, 230–242.

— (1952*b*). Disappearance of histaminase from the whole body following adrenalectomy in cats. *Acta physiol. scand.* **25**, 255–258.

Haeger, K., Kahlson, G. & Westling, H. (1953). Evidence of a regulatory mechanism controlling the levels of histamine and histaminase in the gastro-intestinal tract. *Acta physiol. scand.* **30**, suppl. 111, 177–191.

Hagen, P., Weiner, N., Ono, S. & Lee, F. (1960). Amino acid decarboxylases of mouse mastocytoma tissue. *J. Pharmac. exp. Ther.* **130**, 9–12.

Hagerman, D. D. (1964). Enzymatic capabilities of the placenta. *Fedn Proc.* **23**, 785–790.

Hahn, F., Schmutzler, W., Seseke, G., Giertz, H. & Bernauer, W. (1966). Histaminasefreisetzung durch Heparin und Protamin beim Meerschweinchen. *Biochem. Pharmac.* **15**, 155–160.

Hallenbeck, G. A. & Code, C. F. (1962). Histidine decarboxylase activity in the Walker carcinosarcoma 256. *Proc. Soc. exp. Biol. Med.* **110**, 649–651.

Halpern, B. N., Neveu, T. & Wilson, C. W. M. (1959). The distribution and fate of radioactive histamine in the rat. *J. Physiol.* **147**, 437–449.

Hanson, M. E., Grossman, M. I. & Ivy, A. C. (1948). Doses of histamine producing minimal and maximal gastric secretory responses in dog and man. *Am. J. Physiol.* **153**, 242–258.

Hansson, R., Holmberg, C. G., Tibbling, G., Tryding, N., Westling, H. & Wetterqvist, H. (1966). Heparin-induced diamine oxidase increase in human blood plasma. *Acta med. scand.* **180**, 533–536.

Hansson, R., Tryding, N. & Törnqvist, A. (1969). Diamine oxidase in human pregnancy. The activity of the enzyme in lymph and in blood plasma and the effect of heparin on the latter. *Acta obstet. gynec. scand.* **48**, 8–18.

Hansson, R. & Thysell, H. (1970). The effect of heparin and some related agents on diamine oxidase in rabbit blood plasma. *Acta physiol. scand.* **78**, 539–546.

Harper, A. A., Reed, J. D. & Smy, J. R. (1968). Gastric blood flow in anaesthetized cats. *J. Physiol.* **194**, 795–807.

Harris, G. W. (1955). *Neural Control of the Pituitary Gland.* London: Edward Arnold.

Harris, G. W., Jacobsohn, D. & Kahlson, G. (1952). The occurrence of histamine in cerebral regions related to the hypophysis. *Ciba Foundation Colloquia on Endocrinology* **4**, 186–194.

Hartman, W. J., Clark, W. G. & Cyr, S. D. (1961). Histidine decarboxylase activity of basophils from chronic myelogenous leukemia patients. Origin of blood histamine. *Proc. Soc. exp. Biol. Med.* **107**, 123–125.

Hauschka, T. S., Grinnell, S. T., Révész, L. & Klein, G. (1957). Quantitative studies on the multiplication of neoplastic cells *in vivo*—IV. Influence of doubled chromosome number on growth rate and final population size. *J. nat. Cancer Inst.* **19**, 13–28.

Haverback, B. J., Tecimer, L. B , Dyce, B. J., Cohen, M., Stubrin, M. I. & Santa Ana, A. D. (1964). The effect of gastrin on stomach histamine in the rat. *Life Sci. Oxford* **3**, 637–649.

Heisler, S. & Kovacs, E. M. (1967). The effect of cortisone on gastric histamine content and gastric secretion in pylorus-ligated guineapigs. *Br. J. Pharmac.* **29**, 329–334.

Hemingway, J. T. (1965). The trigger mechanism in mitosis; evidence from humoral factors in plasma of tumour-bearing rats. *J. Physiol.* **179**, 92–93P.

Hicks, R. (1965). Some effects of corticosteroids on tissue histamine levels in the guinea-pig. *Br. J. Pharmac.* **25**, 664–670.

High, D. P., Shephard, D. M. & Woodcock, B. G. (1965). Urinary histamine excretion in the gastrectomized rat. *Life Sci. Oxford* **4**, 787–795.

Hill, J. M. & Mann, P. J. G. (1964). Further properties of the diamine oxidase of pea seedlings. *Biochem. J.* **91**, 171–182.

Hilton, S. M. (1962). Local mechanisms regulating peripheral blood flow. *Physiol. Rev.* **42**, suppl. 5, 265–275.

Hilton, S. M. & Lewis, G. P. (1955). The mechanism of the functional hyperaemia in the submandibular salivary gland. *J. Physiol.* **129**, 253–271.

— (1957). Functional vasodilatation in the submandibular salivary gland. *Br. med. Bull.* **13**, 189–196.

Hilton, S. M. & Torres, S. H. (1970). Selective hypersensitivity to bradykinin in salivary gland with ligated ducts. *J. Physiol.* **211**, 37–48.

Hirschowitz, B. I. & Sachs, G. (1965). Vagal gastric secretory stimulation by 2-deoxy-D-glucose. *Am. J. Physiol.* **209**, 452–460.

Hollander, F. (1952). Current views on the physiology of the gastric secretions. *Am. J. Med.* **13**, 453–464.

Hollander, F. & Schapira, A. (1963). Gastric response to topical histamine; subcutaneous dose–response relation. *Am. J. Physiol.* **205**, 625–630.

Holz, P. & Heise, R. (1937). Uber Histaminbildung im Organismus. *Arch. exp. Path. Pharm.* **186**, 377–386.

Holtz, P. & Palm, D. (1964). Pharmacological aspects of vitamin B_6. *Pharmac. Rev.* **16**, 113–178.

Holtz, P. & Westermann, E. (1956). Uber die Dopadecarboxylase und Histidindecarboxylase des Nervengewebes. *Arch. exp. Path. Pharm.* **227**, 538–546.

Håkanson, R. (1961). Formation of histamine in transplants from a rat mammary carcinoma. *Experientia* **17**, 402.

Håkanson, R. (1964). Histidine decarboxylase in the bone marrow of the rat. *Experientia* **20**, 205–206.

— (1966). Histidine decarboxylase in experimental tumours. *J. Pharm. Pharmac.* **18**, 769–774.

Håkanson, R., Lilja, B. & Owman, C. (1967). Properties of a new system of amine-storing cells in the gastric mucosa of the rat. *Europ. J. Pharmac.* **1**, 188–199.

Håkanson, R. & Owman, C. (1967). Concomitant histochemical demonstration of histamine and catecholamines in enterochromaffin-like cells of gastric mucosa. *Life Sci. Oxford* **6**, 759–766.

Irvine, W. T., Ritchie, H. D. & Adam, H. M. (1961). Histamine concentrations in the gastric venous effluent before and during acid secretion. *Gastroenterology* **41**, 258–263.

Ishikawa, E., Toki, A., Moriyama, T., Matsuoka, Y., Aikawa, T. & Suda, M. (1970). A study on the induction of histidine decarboxylase in tumour-bearing rat. *J. Biochem., Tokyo*, **68**, 347–358.

Ivy, A. C. & Bachrach, W. H. (1966). Physiological significance of the effect of histamine on gastric secretion. In *Handbook of Experimental Pharmacology*, edited by Eichler, O. and Farah, A., vol. 18/1, pp. 810–891, Berlin: Springer.

Jacob, F. & Monod, J. (1961). Genetic regulatory mechanisms in the synthesis of proteins. *J. molec. Biol.* **3**, 318–356.

Jacobson, E. D. (1967). The circulation of the gastrointestinal tract. *Gastroenterology* **52**, 98–112.

— (1968). The gastrointestinal circulation. *A. Rev. Physiol.* **30**, 133–146.

Jacobson, E. D., Swan, K. G. & Grossman, M. I. (1967). Blood flow and secretion in the stomach. *Gastroenterology* **52**, 414–420.

Janowitz, H. & Grossman, M. I. (1949). Minimal effective dose of intravenously administered histamine in pregnant and non-pregnant human beings. *Am. J. Physiol.* **157**, 94–98.

Jepson, K., Duthie, H. L., Fawcett, A. N., Gumpert, J. R., Johnston, D., Lari, J. & Wormsley, K. G. (1968). Acid and pepsin response to gastrin I, pentagastrin, tetragastrin, histamine, and pentagastrin snuff. *Lancet* **ii**, 139–141.

Johansson, M. B. & Wetterqvist, H. (1963). Increased histamine formation in tissues from burned mice. *Med. exp.* **8**, 251–255.

Johnson, H. L. (1969). Nonmast-cell histamine kinetics–II. Effect of histidine decarboxylase inhibitors on rates of decline of tissue ^{3}H-histamine in the female rat. *Biochem. Pharmac.* **18**, 651–658.

Johnson, L. R., Jones, R. S., Aures, D. & Håkanson, R. (1969). Effect of antrectomy on gastric histidine decarboxylase activity in the rat. *Am. J. Physiol.* **216**, 1051–1053.

Johnson, A. R & Moran, N. C. (1967). Comparison of several methods for isolation of rat peritoneal mast cells. *Proc. Soc. exp. Biol. Med.* **123**, 886–889.

Johnson, L. R. & Tumpson, D. B. (1970). Effect of secretin on histamine-stimulated secretion in the gastric fistula rat. *Proc. Soc. exp. Biol. Med.* **133**, 125–127.

Johnston, M. (1967). Histamine formation in rats bearing the Walker mammary carcinosarcoma. *Experientia* **23**, 152–154.

Johnston, M. & Kahlson, G. (1967). Experiments on the inhibition of histamine formation in the rat. *Br. J. Pharmac.* **30**, 274–282.

Jonson, B. & White, T. (1964). Histamine metabolism in the brain of conscious cats. *Proc. Soc. exp. Biol. Med.* **115**, 874–876.

Juhlin, L. (1967). Determination of histamine in small biopsies and histochemical sections. *Acta physiol. scand.* **71**, 30–36.

Juhlin, L. & Shelley, W. B. (1966). Detection of histamine by a new fluorescent *o*-phtalaldehyde stain. *J. Histochem. Cyctochem.* **14**, 525–528.

Jänne, J. (1967). Studies on the biosynthetic pathway of polyamines in rat liver. *Acta physiol. scand.* suppl. 300, 1–71.

Jänne, J. & Raina, A. (1968). Stimulation of spermidine synthesis in the regenerating rat liver: Relation to increased ornithine decarboxylase activity. *Acta chem. scand.* **22**, 1349–1351.

Kahlson, G. (1960). A place for histamine in normal physiology. *Lancet* **i**, 67–71.

— (1962). Nascent histamine and methods of its estimation. *Proc. int. Union physiol. Sci. XXII Physiolog. Congr. Leiden*, vol. 1, pp. 856–862.

Kahlson, G., Lilja, B. & Svensson, S. E. (1964). Physiological protection against gastric ulceration during pregnancy and lactation in the rat. *Lancet* **ii**, 1269–1272.

Kahlson, G., Lindell, S. E. & Westling, H. (1953). The influence of adrenocortical steroids on the histaminase concentration in some organs of the guinea pig. *Acta physiol. scand.* **30**, suppl. 111, 192–201.

Kahlson, G., Nilsson, K., Rosengren, E. & Zederfeldt, B. (1960). Wound healing as dependent on rate of histamine formation. *Lancet* **ii**, 230–234.

Kahlson, G. & Rosengren, E. (1959). Inhibition of histamine formation and some of its consequences. *J. Physiol.* **149**, 66–67P.

— (1960). Retardation of rapid growth in plant tissues (yellow peas) by the enzyme inhibitor semicarbazide. *J. Physiol.* **154**, 2–3P.

— (1963). Histamine formation in bone marrow. *Experientia* **19**, 182–183.

— (1968) New approaches to the physiology of histamine. *Physiol. Rev.* **48**, 155–196.

— (1970). Histamine formation as related to growth and protein synthesis. In *Biogenic Amines as Physiological Regulators*, edited by Blum, J. J., pp. 223–238. Englewood Cliffs, N. J.: Prentice-Hall.

Kahlson, G., Rosengren, E. & Steinhardt, C. (1963a). Histamine-forming capacity of multiplying cells. *J. Physiol.* **169**, 487–498.

— (1963b). Histamine formation in human wound tissue. *Experientia* **19**, 243–244.

Kahlson, G., Rosengren, E., Svahn, D. & Thunberg, R. (1964). Mobilization and formation of histamine in the gastric mucosa as related to acid secretion. *J. Physiol.* **174**, 400–416.

Kahlson, G., Rosengren, E. & Svensson, S. E. (1962). Inhibition of histamine formation *in vivo*. *Nature, Lond.* **194**, 876.

— (1968). Mode of action of a gastric-secretion antagonist. *Br. J. Pharmac.* **33**, 493–500.

Kahlson, G., Rosengren, E. & Thunberg, R. (1963). Observations on the inhibition of histamine formation. *J. Physiol.* **169**, 467–486.

— (1966). Accelerated histamine formation in hypersensitivity reactions. *Lancet* **i**, 782–784.

— (1967). Accelerated mobilization and formation of histamine in the gastric mucosa evoked by vagal excitation. *J. Physiol.* **190**, 455–463.

Kahlson, G., Rosengren, E. & Westling, H. (1958). Increased formation of histamine in the pregnant rat. *J. Physiol.* **143**, 91–103.

Kahlson, G., Rosengren, E., Westling, H. & White, T. (1958). The site of increased formation of histamine in the pregnant rat. *J. Physiol.* **144**, 337–348.

Kahlson, G., Rosengren, E. & White, T. (1959). Formation of histamine by the foetus in the rat and man. *J. Physiol.* **145**, 30–31P.

— (1960). The formation of histamine in the rat foetus. *J. Physiol.* **151**, 131–138.

Kameswaran, L. & West, G. B. (1962). The formation of histamine in mammals. *J. Physiol.* **160**, 564–571.

Kapeller-Adler, R. (1965). Histamine catabolism *in vitro* and *in vivo*. *Fedn Proc.* **24**, 757–765.

— (1970). Amine Oxidases and Methods for their Study. New York and London: John Wiley.

Karady, S., Rose, B. & Browne, J. S. L. (1940). Decrease of histaminase in tissue by adrenalectomy and its restoration by cortico-adrenal extract. *Am. J. Physiol.* **130**, 539–542.

Keele, C. A. & Armstrong, D. (1964). *Substances Producing Pain and Itch*. London: Edward Arnold.

Kelly, P. & Robert, A. (1969). Inhibition by pregnancy and lactation of steroid-induced ulcers in the rat. *Gastroenterology* **56**, 24–29.

Kim, Y. S. & Glick, D. (1967). Determination of histidine decarboxylase in microgram samples of tissue, properties and quantitative histochemistry of the enzyme and histamine in the rat stomach. *J. Histochem. Cytochem.* **15**, 347–352.

— (1968). Quantitative histochemistry of histamine and histidine decarboxylase in the stomach of rat, pig, and dog in various functional states. *Gastroenterology* **55**, 657–664.

Kim, K. S., Ridley, P. T. & Tuegel, C. (1968). Effect of insulin on gastric acid secretion, histamine formation, and on the incidence of gastric lesions. *Life Sci. Oxford* **7**, 403–409.

Kobayashi, Y. (1963). Determination of histidine decarboxylase activity by liquid scintillation counting of $C^{14}O_2$. *Analyt. Biochem.* **5**, 284–290.

— (1964). Plasma diamine oxidase titres of normal and pregnant rats. *Nature, Lond.* **203**, 146–147.

Komarov, S. A. (1942). Studies on gastrin—I. Methods of isolation of a specific gastric secretagogue from the pyloric mucous membrane and its chemical properties. *Rev. can. Biol.* **1,** 191–205.

Kowalewski, K., Russell, J. C. & Koheil, A. (1969). Blood and tissue histamine in mice, rats, hamsters, and guinea pigs. *Proc. Soc. exp. Biol. Med.* **132,** 443–446.

Kullander, S. (1952). Studies on eosinophil leucocytes in pregnancy. *Acta endocr., Copenh.* **10,** 135–148.

Kutscher, F. (1910). Die physiologische Wirkung einer Secalebase und des Imidazolyläthylamins. *Zentbl. Physiol.* **24,** 163–165.

Kwiatkowski, H. (1943). Histamine in nervous tissue. *J. Physiol.* **102,** 32–41.

Lajtha, L. G. (1957). Bone marrow cell metabolism. *Physiol. Rev.* **37,** 50–65.

LeBlond, C. P. & Walker, B. E. (1956). Renewal of cell population. *Physiol. Rev.* **36,** 255–276.

Lecomte, J. (1961). Sur les facteurs surrénaliens qui conditionnent la résistance du rat intoxiué par l'histamine. *Arch. int. Physiol. Biochem.* **69,** 563–581.

Leitch, J. L., Debley, V. G. & Haley, T. J. (1956). Endogenous histamine excretion in the rat as influenced by X-ray irradiation and compound 48/80. *Am. J. Physiol.* **187,** 307–311.

Levine, R. J. (1965). Effect of histidine decarboxylase inhibition on gastric acid secretion in the rat. *Fedn. Proc.* **24,** 1331–1333.

— (1966). Histamine synthesis in man: inhibition by 4-bromo-3-hydroxybenzyloxyamine. *Science, N.Y.* **154,** 1017–1019.

— (1968). Histamine and the effect of gastrin on gastric acid secretion in man. *Fedn. Proc.* **27,** 1341–1344.

Levine, R. J., Sato, T. L. & Sjoerdsma, A. (1965). Inhibition of histamine synthesis in the rat by α-hydrazino analog of histidine and 4-bromo-3-hydroxybenzyloxyamine. *Biochem. Pharmac.* **14,** 139–149.

Levine, R. J. & Watts, D. E. (1966). A sensitive and specific assay for histidine decarboxylase activity. *Biochem. Pharmac.* **15,** 841–849.

Lewis, T. (1927). *The Blood Vessels of the Human Skin and Their Responses.* London: Shaw.

Lilja, B. & Lindell, S. E. (1961). Metabolism of ^{14}C-histamine in heart-lung-liver preparations of cats. *Br. J. Pharmac.* **16,** 203–208.

Lilja, B., Lindell, S. E. & Saldeen, T. (1960). Formation and destruction of ^{14}C-histamine in human lung tissue *in vitro. J. Allergy,* **31,** 492–496.

Lilja, B. & Svensson, S. E. (1967). Gastric secretion during pregnancy and lactation in the rat. *J. Physiol.* **190,** 261–272.

Lindahl, K. M. (1958). Methylation of histamine by mouse liver *in vitro. Acta physiol. scand.* **43,** 254–261.

— (1960). The histamine methylation enzyme system in liver. *Acta physiol. scand.* **49,** 114–138.

Lindberg, S. (1963*a*). ^{14}C-histamine elimination from the blood of pregnant and non-pregnant women with special reference to the uterus. *Acta obstet. gynec. scand.* **42,** suppl. 1, 3–25.

Lindberg, S. (1963*b*). ¹⁴C-histamine inactivation *in vitro* by human myome-trial and placental tissues. *Acta obstet. gynec. cand.* **42**, suppl. 1, 26–34.

Lindberg, S., Lindell, S. E. & Westling, H. (1963*a*). The metabolism of ¹⁴C-labelled histamine injected into the umbilical artery. *Acta obstet. gynec. scand.* **42**, suppl. 1, 35–47.

— (1963*b*). Formation and inactivation of histamine by human foetal tissues *in vitro*. *Acta obstet. gynec. scand.* **42**, suppl. 1, 49–58.

Lindberg, S. & Törnqvist, Å. (1966). The inhibitory effect of amino-guanidine on histamine catabolism in human pregnancy. *Acta obstet. gynec. scand.* **45**, 131–139.

Lindell, S. E. (1957). Effects of histaminase inhibitors on the blood pressure responses in dogs to histamine injected into the renal artery. *Acta physiol. scand.* **41**, 168–186.

Lindell, S. E., Rorsman, H. & Westling, H. (1959). Formation of hista-mine in a canine mastocytoma. *Experientia* **15**, 31.

— (1961*a*). Histamine formation in urticaria pigmentosa. *Acta derm. venerol.* **41**, 277–280.

— (1961*b*). Histamine formation in human blood. *Acta allerg.* **16**, 216–227.

Lindell, S. E. & Schayer, R. W. (1958). Metabolism of injected C¹⁴-histamine in the kidney of the dog. *Br. J. Pharmac.* **13**, 52–54.

Lindell, S. E. & Westling, H. (1953). The histaminase activity of different organs in the guinea-pig. *Acta physiol. scand.* **30**, suppl. 111, 202–206.

— (1954). Potentiation of histamine effects by an antihistaminase. *Acta physiol. scand.* **32**, 230–237.

— (1956). Potentiation by histaminase inhibitors of the blood pressure responses to histamine in cats. *Acta physiol. scand.* **37**, 307–323.

— (1962). Blood concentration and biological effects of infused C¹⁴-histamine in man. *Proc. int. Union Physiol. Sci. XXII Int. Physiolog. Congr. Leiden.* vol. 2, p. 702.

— (1966). Histamine metabolism in man. In *Handbook of Experimental Pharmacology*, edited by Eichler, O. & Farah, A. vol. 18/1, pp. 734–788. Berlin: Springer.

Lindell, S. E., Westling, H. & Zederfeldt, B. (1964). Histamine forming capacity of human wound tissue. *Proc. Soc. exp. Biol. Med.* **116**, 1054–1055.

Lorenz, W., Halbach, S., Gerant, M. & Werle, E. (1969). Specific histi-dine decarboxylases in the gastric mucosa of man and other mammals. *Biochem. Pharmac.* **18**, 2625–2637.

Lorenz, W., Heitland, S., Werle, E., Schauer, A. & Gastpar, P. (1968). Histamin in Speicheldrusen, Tonsillen und Thymus und adaptive Histaminbildung in der Glandula Submaxillaris. *Naunyn-Schmiede-bergs Arch. Pharmak. exp. Path.* **259**, 319–328.

Lovenberg, W., Weissbach, H. & Udenfriend, S. (1962). Aromatic L-amino acid decarboxylase. *J. biol. Chem.* **237**, 89–93.

MacIntosh, F. C. (1938). Histamine as a normal stimulant of gastric secretion. *Q. J. exp. Physiol.* **28**, 87–98.

MacIntosh, F. C. & Paton, W. D. M. (1949). The liberation of histamine by certain organic bases. *J. Physiol.* **109**, 190–219.

306   *Biogenesis and physiology of histamine*

Mackay, D., Marshall, P. B. & Riley, J. F. (1960). Histidine decarboxylase activity in a malignant rat hepatoma. *J. Physiol.* **153**, 31P.

Mackay, D., Reid, J. D. & Shephard, D. M. (1961). Histidine decarboxylase and growth. *Nature, Lond.* **191**, 1311–1312.

Mackay, D. & Shephard, D. M. (1960). A study of potential histidine decarboxylase inhibitors. *Br. J. Pharmac.* **15**, 552–556.

Makhlouf, G. M., McManus, J. P. A. & Card, W. I. (1964). The action of gastrin II on gastric-acid secretion in man. *Lancet*, **ii**, 485–490.

— (1965). A comparative study of the effects of gastrin, histamine, histalog, and mechothane on the secretory capacity of the human stomach in two normal subjects over 20 months. *Gut* **6**, 525–534.

Malkiel, S. & Hargis, B. J. (1952). Histamine sensitivity and anaphylaxis in the pertussis-vaccinated rat. *Proc. Soc. exp. Biol. Med.* **81**, 689–691.

Marcou, I., Athanasiu-Vergu, E., Chiricéanu, D., Cosma, G., Gingold, N. & Parhon, C. C. (1938). Sur le role physiologique de l'histamine. *Presse méd.* **1**, 371–374.

Martin, L. (1962). The effects of histamine on the vaginal epithelium of the mouse. *J. Endocr.* **23**, 329–340.

Maudsley, D. V., Radwan, A. G. & West, G. B. (1967). Comparison of isotopic and non-isotopic methods of estimating histidine decarboxylase activity. *Br. J. Pharmac.* **31**, 313–318.

Mellanby, E. (1916). An experimental investigation on diarrhoea and vomiting of children. *Q. Jl. Med.* **9**, 165–215.

Mellander, S. (1970). Systemic circulation: local control. *A. Rev. Physiol.* **32**, 313–344.

Mellander, S. & Johansson, B. (1968). Control of resistance exchange and capacitance functions in the peripheral circulation. *Pharmac. Rev.* **20**, 117–196.

Misrahy, G. A. (1946). The metabolism of histamine and adenylic compounds in the embryo. *Am. J. Physiol.* **147**, 462–470.

Mitchell, R. G. (1963). Histidine decarboxylase in the new-born human infant. *J. Physiol.* **166**, 136–144.

Mitchell, R. G. & Cass, R. (1959). Histamine and 5-hydroxytryptamine in the blood of infants and children. *J. clin. Invest.* **38**, 595–604.

Mongar, J. L. & Schild, H. O. (1951). Potentiation of the action of histamine by semicarbazide. *Nature, Lond.* **167**, 232–233.

— (1952). A comparison of the effects of anaphylactic shock and of chemical histamine releasers. *J. Physiol.* **118**, 461–478.

— (1957). Inhibition of the anaphylactic reaction. *J. Physiol.* **135**, 301–319.

— (1962). Cellular mechanisms in anaphylaxis. *Physiol. Rev.* **42**, 226–270.

Mongar, J. L. & Whelan, R. F. (1953). Histamine release by adrenaline and D-tubocurarine in the human subject. *J. Physiol.* **120**, 146–154.

Moody, F. G. (1967). Gastric blood flow and acid secretion during intraarterial histamine administration. *Gastroenterology* **52**, 216–224.

Moolten, F. L. & Bucher, N. L. R. (1967). Regeneration of rat liver: transfer of humoral agent by cross circulation. *Science, N.Y.* **158**, 272–274.

Morley, J. S., Tracy, H. J. & Gregory, R. A. (1965). Structure-function relationships in the active C-terminal tetrapeptide sequence of gastrin. *Nature, Lond.* **207**, 1356–1359.

Mota, I. (1966). Release of histamine from mast cells. In *Handbook of Experimental Pharmacology*, edited by Eichler, O. & Farah, A. vol. 18/1, pp. 569–636. Berlin: Springer.

Moyed, H. S. & Umbarger, H. E. (1962). Regulation of biosynthetic pathways. *Physiol. Rev.* **42**, 444–466.

Myren, J., Unhjem, O., Gjeruldsen, S. T. & Semb, L. S. (1965). Oxygen consumption of unstimulated gastric mucosa in relation to the secretion of hydrochloric acid and the number of parietal cells. *Scand. J. clin. Lab. Invest.* **17**, 31–38.

Newton, W. H. (1949). *Recent Advances in Physiology* (7th edition), pp. 86–112. London: Churchill.

Niedzielski, A. & Maśliński, C. (1968). Histidine decarboxylase in mouse foetal tissues. 1. The characteristics of the foetal enzyme. *Europ. J. Pharmac.* **4**, 457–463.

Nilsson, K., Lindell, S. E., Schayer, R. W. & Westling, H. (1959). Metabolism of ^{14}C-labelled histamine in pregnant and non-pregnant women. *Clin. Sci.* **18**, 313–319.

Noll, W. W. & Levine, R. J. (1970). Histidine decarboxylase in gastric tissues of primates. *Biochem. Pharmac.* **19**, 1043–1053.

Nylander, G. & Olerud, S. (1961). The vascular pattern of the gastric mucosa of the rat following vagotomy. *Surgery, Gynec. Obstet.* **112**, 475–480.

Öbrink, K. J. (1948). Studies on the kinetics of the parietal secretion of the stomach. *Acta physiol. scand.* **15**, suppl. 51, 1–106.

Ono, S. & Hagen, P. (1959). Pyridoxal phosphate: a coenzyme for histidine decarboxylase. *Nature, Lond.* **184**, 1143–1144.

Parratt, J. R. & West, G. B. (1960). Hypersensitivity and the thyroid gland. *Int. Archs Allergy appl. Immun.* **16**, 288–302.

Paton, W. D. M. & Vane, J. R. (1963). An analysis of the responses of the isolated stomach to electrical stimulation and to drugs. *J. Physiol.* **165**, 10–46.

Pearlman, D. S. & Waton, N. G. (1966). Observations on the histamine forming capacity of mouse tissues and of its potentiation after adrenaline. *J. Physiol.* **183**, 257–268.

Peters, J. H. & Gordon, G. R. (1968). Histaminase activities in the plasma of subhuman primates and man. *Nature, Lond.* **217**, 274–275.

Raina, A. & Jänne, J. (1970). Polyamines and the accumulation of RNA in mammalian systems. *Fedn Proc.* **29**, 1568–1574.

Raina, A., Jänne, J. & Siimes, M. (1965). Polyamines and nucleic acids in regenerating rat liver. In *Abstr. 2nd Meeting Fed. Europ. Biochem. Soc.* (*A*114) *Vienna*.

— (1966). Stimulation of polyamine synthesis in relation to nucleic acids in regenerating rat liver. *Biochim. biophys. Acta* **123**, 197–201.

Reid, J. D. & Shephard, D. M. (1963). Inhibition of histidine decarboxylase. *Life Sci., Oxford* **2**, 5–8.

Reilly, M. A. & Schayer, R. W. (1968). Further studies on the histidine–histamine relationship *in vivo*: effects of endotoxin and of histidine decarboxylase inhibitors. *Br. J. Pharmac.* **34**, 551–563.

— (1970). *In vivo* studies on histamine catabolism and its inhibition. *Br. J. Pharmac.* **38**, 478–489.

Reite, O. B. (1965). A phylogenetical approach to the functional significance of tissue mast cell histamine. *Nature, Lond.* **206**, 1334–1336.

Riley, J. F. (1959). *The Mast Cells*. Edinburgh and London: E. & S. Livingstone.

Riley, J. F. & West, G. B. (1953). The presence of histamine in tissue mast cells. *J. Physiol.* **120**, 528–537.

— (1966). The occurrence of histamine in mast cells. In *Handbook of Experimental Pharmacology*, edited by Eichler, O. & Farah, A. vol. 18/1, pp. 116–135. Berlin: Springer.

Roberts, M. & Adam, H. M. (1950). New methods for the quantitative estimation of free and conjugated histamine in body fluids. *Br. J. Pharmac.* **5**, 526–541.

Roberts, M. & Robson, J. M. (1953). The histaminase content of the rat uterus, and its relation to the decidua. *J. Physiol.* **119**, 286–291.

Robinson, B. & Shephard, D. M. (1961a). The preparation of DL-α-methyl-histidine dihydrochloride. *J. chem. Soc.* 5037–5038.

— (1961b). Inhibition of histidine decarboxylases. *Biochim. biophys. Acta* **53**, 431–433.

— (1962). The inhibition of the L-histidine decarboxylases of guinea-pig kidney and rat hepatoma. *J. Pharm. Pharmac.* **14**, 9–15.

Rocha e Silva, M. (1966). Release of histamine in anaphylaxis. In *Handbook of Experimental Pharmacology*, edited by Eichler, O. & Farah, A. vol. 18/1, pp. 431–480. Berlin: Springer.

Rose, B. (1939). The effect of cortin and desoxycorticosterone acetate on the ability of the adrenalectomized rat to inactivate histamine. *Am. J. Physiol.* **127**, 780–784.

Rose, B. & Browne, J. S. L. (1938). The distribution and rate of disappearance of intravenously injected histamine in the rat. *Am. J. Physiol.* **124**, 412–420.

Rosengren, E. (1962). Formation of histamine in the pregnant mouse. *Experientia* **18**, 176–177.

— (1963). Histamine metabolism in the pregnant mouse. *J. Physiol.* **169**, 499–512.

— (1965). Histamine metabolism in the pregnant and non-pregnant hamster. *Proc. Soc. exp. Biol. Med.* **118**, 884–888.

— (1966). Histamine metabolism in pregnancy. *Acta Univ. Lundensis* sec. II, no. 8, 1–26.

Rosengren, E. & Steinhardt, C. (1961). Elevated histidine decarboxylase activity in the kidney of the pregnant mouse. *Experientia* **17**, 544.

Rosengren, E. & Svensson, S. E. (1969). The role of the antrum and the vagus nerve in the formation of gastric mucosal histamine. *J. Physiol.* **205**, 275–288.

— (1970). Gastrin derivatives investigated for secretory potency and for changes in gastric mucosal histamine formation. *Br. J. Pharmac.* **38,** 473–477.

Rothschild, A. M. & Schayer, R. W. (1958a). Histidine decarboxylase from rat peritoneal fluid mast cells. *Fedn Proc.* **17,** 136.

— (1958b). Synthesis and metabolism of a histamine metabolite, 1-methyl-4-(β-aminoethyl)-imidazole. *Biochim. biophys. Acta* **30,** 23–27.

— (1959). Characterization of histidine decarboxylase from rat peritoneal fluid mast cells. *Biochim. biophys. Acta* **34,** 392–398.

Roy, C. S. & Brown, J. G. (1879–80). The blood pressure and its variations in the arterioles, capillaries and smaller veins. *J. Physiol.* **2,** 323–359.

Russell, B. (1968). *The Autobiography of Bertrand Russell.* Vol. 2. London: George Allen and Unwin.

Russell, D. & Snyder, S. H. (1968). Amine synthesis in rapidly growing tissues: ornithine decarboxylase activity in regenerating rat liver, chick embryo, and various tumours. *Proc. natn. Acad. Sci. U.S.A.* **60,** 1420–1427.

Räsänen, T. (1958). Tissue eosinophils and mast cells in the human stomach wall in normal and pathological conditions. *Acta path. microbiol. scand.* suppl. 129, 1–131.

Sandberg, N. (1964a). Enhanced rate of healing in rats with an increased rate of histamine formation. *Acta chir. scand.* **127,** 9–21.

— (1964b). Time relationship between administration of cortisone and wound healing in rats. *Acta chir. scand.* **127,** 446–455.

Sandberg, N. & Steinhardt, C. (1964). On the effect of cortisone on the histamine formation of skin wounds in the rat. *Acta chir. scand.* **127,** 574–577.

Sandblom, P., Petersen, P. & Muren, A. (1953). Determination of the tensile strength of the healing wound as a clinical test. *Acta chir. scand.* **105,** 252–257.

Schachter, M. (1953). Anaphylaxis and histamine release in the rabbit. *Br. J. Pharmac.* **8,** 412–419.

Schachter, M. & Talesnik, J. (1952). The release of histamine by egg-white in non-sensitized animals. *J. Physiol.* **118,** 258–263.

Schayer, R. W. (1952). Biogenesis of histamine. *J. biol. Chem.* **199,** 245–250.

— (1956a). Formation and binding of histamine by free mast cells of rat peritoneal fluid. *Am. J. Physiol.* **186,** 199–202.

— (1956b). Formation and binding of histamine by rat tissues *in vitro.* *Am. J. Physiol.* **187,** 63–65.

— (1957). Histidine decarboxylase of rat stomach and other mammalian tissues. *Am. J. Physiol.* **189,** 533–536.

— (1959). Catabolism of physiological quantities of histamine *in vivo.* *Physiol. Rev.* **39,** 116–126.

— (1960). Relationship of stress-induced histidine decarboxylase to circulatory homeostasis and shock. *Science, N.Y.* **131,** 226–227.

Schayer, R. W. (1962a). Role of induced histamine in tourniquet shock in mice. *Am. J. Physiol.* **203**, 412–416.

— (1962b). Evidence that induced histamine is an intrinsic regulator of the microcirculatory system. *Am. J. Physiol.* **202**, 66–72.

— (1966a). Enzymatic formation of histamine from histidine. In *Handbook of Experimental Pharmacology*, edited by Eichler, O. & Farah, A. vol. 18/1, pp. 688–725. Berlin: Springer.

— (1966b). Catabolism of histamine *in vivo*. In *Handbook of Experimental Pharmacology*, edited by Eichler, O. & Farah, A. vol. 18/1, pp. 672–683. Berlin: Springer.

— (1967). A unified theory of glucocorticoid action—II. On a circulatory basis for the metabolic effects of glucocorticoids. *Perspect. Biol. Med.* **10**, 409–418.

— (1968a) Determination of histidine decarboxylase activity. *Methods of Biochemical Analysis.* **16**, 273–291.

— (1968b). Histamine and a possible unity of autonomous microcirculatory dilator responses. *MCV Quarterly* **4** (3), 101–106.

Schayer, R. W. & Cooper, J. A. D. (1956). Metabolism of C^{14}-histamine in man. *J. appl. Physiol.* **9**, 481–483.

Schayer, R. W., Davis, K. J. & Smiley, R. L. (1955). Binding of histamine *in vitro* and its inhibition by cortisone. *Am. J. Physiol.* **182**, 54–56.

Schayer, R. W. & Ganley, O. H. (1959). Adaptive increase in mammalian histidine decarboxylase activity in response to nonspecific stress. *Am. J. Physiol.* **197**, 721–724.

— (1961). Relationship of increased histidine decarboxylase activity to *Bordetella pertussis* vaccine sensitization of mice. *J. Allergy* **32**, 204–213.

Schayer, R. W. & Ivy, A. C. (1957). Evidence that histamine is a gastric secretory hormone in the rat. *Am. J. Physiol.* **189**, 369–372.

— (1958). Release of C^{14}-histamine from stomach and intestine on feeding. *Am. J. Physiol.* **193**, 400–402.

Schayer, R. W. & Reilly, M. A. (1968). Suppression of inflammation and histidine decarboxylase by protein synthesis inhibitors. *Am. J. Physiol.* **215**, 472–476.

Schayer, R. W., Rothschild, Z. & Bizony, P. (1959). Increase in histidine decarboxylase activity of rat skin following treatment with compound 48/80. *Am. J. Physiol.* **196**, 295–298.

Schayer, R. W. & Sestokas, E. (1965). Suppression of histamine synthesis in guinea-pigs by α-methyl-3,4-dihydroxyphenylalanine. *Biochim. biophys. Acta* **111**, 557–558.

Schayer, R. W., Smiley, R. L. & Davis, K. J. (1954). Inhibition by cortisone of the binding of new histamine in rat tissues. *Proc. Soc. exp. Biol. Med.* **87**, 590–592.

Schayer, R. W., Smiley, R. L. & Kennedy, J. (1952). Effect of adrenalectomy on rate of metabolism of histamine in the mouse. *Proc. Soc. exp. Biol. Med.* **81**, 416–417.

Schild, H. O., Hawkins, D. F., Mongar, J. L. & Herxheimer, H. (1951).

Reactions of isolated human asthmatic lung and bronchial tissue to a specific antigen. *Lancet* **ii**, 376–381.

Schmitterlöw, C. G. (1948). The nature and occurrence of pressor and depressor substances in extracts from blood vessels. *Acta physiol. scand.* **16**, suppl. 56, 1–113.

Schmutzler, W., Bethge, K. P. & Moritz, G. (1967). Release of histaminase and its resynthesis in the guinea-pig liver. *Naunyn-Schmiedebergs Arch. Pharmak. Exp. Path.* **259**, 192.

Shay, H., Sun, D. C. H. & Gruenstein, M. (1954). A quantitative method for measuring spontaneous gastric secretion in the rat. *Gastroenterology* **26**, 906–913.

Shephard, D. M. & Woodcock, B. G. (1968). Histamine formation in normal, regenerating and malignant liver tissue of the rat. *Biochem. Pharmac.* **17**, 23–30.

Shore, P. A., Burkhalter, A. & Cohn, V. H. (1959). A method for the fluoremetric assay of histamine in tissues. *J. Pharmac. exp. Ther.* **127**, 182–186.

Siesjö, B. (1962). The solubility of carbon dioxide in cerebral cortical tissue from cats at 37·5°C. *Acta physiol. scand.* **55**, 325–341.

Sjaastad, O. (1967). Possible *in vivo* de-acetylation of N-acetylhistamine in man. *Nature, Lond.* **216**, 1111–1112.

Sjaastad, O. & Nygaard, K. (1967). Urinary and faceal histamine in rats with a blind jejunal pouch. *Med. Pharmac. exp.* **17**, 273–280.

Smith, A. N. (1954). Gastrin and histamine release. *J. Physiol.* **123**, 71–72P.

— (1959). The distribution and release of histamine in human gastric tissues. *Clin. Sci.* **18**, 533–541.

— (1961). Histamine and gastric secretion. *J. R. Coll. Surg. Edinb.* **6**, 276–292.

Smith, J. K. (1967). The purification and properties of placental histaminase. *Biochem. J.* **103**, 110–119.

Smith, R. D. & Code, C. F. (1967). Histamine formation: histidine decarboxylase determination using carboxyl-^{14}C-labelled histidine. *Mayo clin. Proc.* **42**, 105–111.

Smith, S. E. (1960). The pharmacological actions of 3,4-di-hydroxyphenyl-α-methylalanine (α-methyl DOPA), an inhibitor of 5-hydroxytryptophan decarboxylase. *Br. J. Pharmac.* **15**, 319–327.

Smyth, R. D., Lambert, R. & Martin, G. J. (1964). Quercetin inhibition of specific histidine decarboxylase. *Proc. Soc. exp. Biol. Med.* **116**, 593–596.

Snyder, S. H., Axelrod, J. & Bauer, H. (1964). The fate of C[14]-histamine in animal tissues. *J. Pharmac. exp. Ther.* **144**, 373–379.

Snyder, S. H. & Epps, L. (1968). Regulation of histidine decarboxylase in rat stomach by gastrin: the effect of inhibitors of protein synthesis. *Mol. Pharmac.* **4**, 187–195.

Staszewska-Barczak, J. & Vane, J. R. (1965). The release of catechol amines from the adrenal medulla by histamine. *Br. J. Pharmac.* **25**, 728–742.

Stevens, C. E. & Leblond, C. P. (1953). Renewal of the mucous cells in the gastric mucosa of the rat. *Anat. Rec.* **115**, 231–243.

Suzuki, T., Hirai, K., Yoshio, H., Kurouji, K. & Yamashita, K. (1963). Effect of histamine on adrenal 17-hydroxycorticoid secretion in unanaesthetized dogs. *Am. J. Physiol.* **204**, 847–848.

Svensson, S. E. (1969). Gastric secretion in unanaesthetized Pavlov pouch rats. *J. Physiol.* **200**, 116–118P.

— (1970*a*). The secretory pattern of three stomach preparations in the rat. *J. Physiol.* **207**, 329–350.

— (1970*b*). Secretory behaviour and histamine formation in the rat Heidenhain pouch following antrectomy. *J. Physiol.* **207**, 699–708.

Svensson, S. E. & Wetterqvist, H. (1968). Nicotine-induced alterations in histamine metabolism in the rat. *Br. J. Pharmac.* **33**, 570–575.

Swan, K. G. & Jacobson, E. D. (1967). Gastric blood flow and secretion in conscious dogs. *Am. J. Physiol.* **212**, 891–896.

Swanberg, H. (1950). Histaminase in pregnancy with special reference to its origin and formation. *Acta physiol. scand.* **23**, suppl. 79, 1–69.

Szego, C. M. (1965). Role of histamine in mediation of hormone action. *Fedn. Proc.* **24**, 1343–1352.

Tabor, H. (1954). Metabolic studies on histidine, histamine and related imidazoles. *Pharmac. Rev.* **6**, 299–343.

— (1956). The fate of histamine in the body. In *Ciba Foundation Symposium on Histamine*, edited by Wolstenholme, G. E. W., O'Connor, C. M. pp. 318–338. London: Churchill.

Tabor, H. & Tabor, C. W. (1964). Spermidine, spermine, and related amines. *Pharmac. Rev.* **16**, 245–300.

Takaya, K. (1969). The relationship between mast cells and histamine in phylogeny with special reference to reptiles and birds. *Arch. histol. jap.* **30**, 401–420.

Telford, J. M. & West, G. B. (1960). The effects of corticosteroids and related compounds on the histamine and 5-hydroxytryptamine content of rat tissues. *Br. J. Pharmac.* **15**, 532–539.

Tham, R. (1966). Liberation of histamine in man. Gas chromatography of ring methylated imidazoleacetic acids in urine. *Scand. J. clin. Lab. Invest.* **18**, 603–616.

Thayer, W. R. & Martin, H. F. (1967). Histidine decarboxylase inhibition and gastric secretion. *Am. J. dig. Dis.* **12**, 1050–1061.

Thornton, G. H. M. & Clifton, J. A. (1959). Estimation of gastric hydrochloric acid secretion in rats by a test meal technique. *Am. J. Physiol.* **197**, 263–268.

Thunberg, R. (1967). Localization of cells containing and forming histamine in the gastric mucosa of the rat. *Expl. Cell Res.* **47**, 108–115.

Tjio, J. H. & Levan, A. (1954). Chromosome analysis of three hyperdiploid ascites tumours of the mouse. *Kungl. Fysiogr. Sällskap. Handl. N. F.* **65**, No. 15, 1–38

Tracy, H. J. & Gregory, R. A. (1964). Physiological properties of a series of synthetic peptides structurally related to gastrin I. *Nature, Lond.* **204**, 935–938.

Tryding, N. (1965). Heparin-induced diamine oxidase (DAO) activity. *Scand. J. clin. Lab. Invest.* **17**, suppl. 86, 196.

Törnqvist, Å. (1968). Some aspects of the influence of histaminase on the sensitivity to and the metabolism of exogenous histamine during pregnancy. *Acta obstet. gynec. scand.* **47**, suppl. 9.

Udenfriend, S., Clark, C. T. & Titus, E. (1952). The presence of 5-hydroxytryptamine in the venom of *Bufo marinus. Experientia* **8**, 379–380.

Umbreit, W. W. & Waddell, J. G. (1949). Mode of action of desoxypyridoxine. *Proc. Soc. exp. Biol. Med.* **70**, 293–299.

Vogt, M. (1954). The concentration of sympathin in different parts of the central nervous system under normal conditions and after the administration of drugs. *J. Physiol.* **123**, 451–481.

Ward-Orsini, M. (1954). The trophoblastic giant cells and endovascular cells associated with pregnancy in the hamster, *Cricetus Auratus. Am. J. Anat.* **94**, 273–321.

Waton, N. G. (1956). Studies on mammalian histidine decarboxylase. *Br. J Pharmac.* **11**, 119–127.

— (1962). Histamine metabolism in foetal and very young cats. *J. Physiol.* **161**, 12–13P.

— (1963). Uptake of ¹⁴C-histamine by cat tissues. *J. Physiol.* **168**, 6–7P.

— (1964). Uptake and formation of histamine by cat tissues. *J. Physiol.* **172**, 475–481.

Weissbach, H., Lovenberg, W. & Udenfriend, S. (1961). Characteristics of mammalian histidine decarboxylating enzymes. *Biochim. biophys. Acta* **50**, 177–179.

Werle, E. (1936). Uber die Bildung von Histamin aus Histidin durch tierisches Gewebe. *Biochem. Z.* **288**, 292–293.

— (1961). Hemmung der Histidin-Decarboxylase durch α-Methyldopa. *Naturwissenschaften*, **48**, 54–55.

— (1964). Aminoxydasen. *Handbuch der Physiologisch-und Pathologisch-Chemischen Analyse.* Hoppe-Seyler/Thierfelder. vol. 6/A, pp. 653–704. Berlin: Springer.

Werle, E. & Heitzer, K. (1938). Zur Kenntnis der Histidindecarboxylase. *Biochem. Z.* **299**, 420–436.

Werle, E. & Pechmann, E. (1949). Uber die Diamin-oxydase der Pflanzen und ihre adaptive Bildung durch Bakterien. *Justus Liebigs annln Chem.* **562**, 44–60.

Werle, E. & Raub, A. (1948). Uber Vorkommen, Bildung und Abbau biogener Amine bei Pflenzen unter besonderer berüchsicktigung des Histamins. *Biochem. Z.* **318**, 538–553.

Werle, E., Schauer, A. & Buhler, H. W. (1963). Klassifizierung Histidin-decarboxylierende Gewebs- und Bakterienenzyme. *Archs int. Pharmacodyn. Ther.* **145**, 198–206.

Westling, H. (1958). The difference in the metabolism of injected ¹⁴C-histamine in male and female rats. *Br. J. Pharmac.* **13**, 498–500.

Westling, H. & Wetterqvist, H. (1962). Further observations on the difference in the metabolism of histamine in male and female rats. *Br. J. Pharmac.* **19**, 64–73.

Wetterqvist, H. & White, T. (1968). Urinary excretion of histamine and methylhistamine in male and female rats. *Scand. J. clin. Lab. Invest.* **22**, suppl. 104, 3–12.

Whelan, R. F. (1956). Histamine and vasodilatation. In *Ciba Foundation Symposium on Histamine*, edited by Wolstenholme, G. E. W. & O'Connor, C. M. pp. 220–234. London: Churchill.

Whelan, R. F. & de la Lande, I. S. (1963). Action of adrenaline on limb blood vessels. *Brit. med. Bull.* **19**, 125–131.

White, T. (1959). Formation and catabolism of histamine in brain tissue *in vitro. J. Physiol.* **149**, 34–42.

— (1960). Formation and catabolism of histamine in cat brain *in vivo. J. Physiol.* **152**, 299–308.

— (1961). Inhibition of the methylation of histamine in cat brain. *J. Physiol.* **159**, 191–197.

— (1964). Biosynthesis, metabolism, and function of histamine in the nervous system. *Fedn Proc.* **23**, 1103–1106.

— (1966). Histamine and methylhistamine in cat brain and other tissues. *Br. J. Pharmac.* **26**, 494–501.

Wickmann, B. E. (1955). The mast cell count during the process of wound healing. *Acta path. microbiol. scand.* suppl. 108, 1–35.

Wicksell, F. (1949a). Observations on histamine and histaminolysis in pregnancy. *Acta physiol. scand.* **17**, 395–414.

— (1949b). A simplified method for estimating the histaminolytic activity of plasma in pregnancy. *Acta physiol. scand.* **17**, 359–369.

Wilson, C. W. M. (1954). The metabolism of histamine as reflected by changes in its urinary excretion in the rat. *J. Physiol.* **125**, 534–545.

Wiscont-Buczkowska, H. (1961). Relationship between the decrease in urinary excretion of histamine and the initiation of the labour process. *Nature, Lond.* **191**, 290.

Zeller, E. A. (1942). Diamine-oxydase. *Adv. Enzymol.* **2**, 93–112.

— (1963). Diamine oxidases. In *The Enzymes* (2nd ed.) edited by Boyer, P. D., Lardy, H. A. & Myrbäck, K., vol. 8, pp. 313–335. New York: Academic Press.

— (1965). Identity of histaminase and diamine oxidase. *Fedn. Proc.* **24**, 766–768.

Zeller, E. A., Schär, B. & Staehlin, S. (1939). Weitere Beiträge zur Kenntnis der Diamin-oxydase (Histaminase). *Helv. chim. Acta* **22**, 837–850.

SUBJECT INDEX

MAGDALEN LIBRARY COLLEGE